FROM START TO FINISH

FROM START TO FINISH

A PRACTICAL GUIDE TO BECOMING A SCIENTIST IN PSYCHOLOGY AND NEUROSCIENCE

Ashleigh M. Maxcey, PhD
The Ohio State University

Geoffrey F. Woodman, PhD
Vanderbilt University

Bassim Hamadeh, CEO and Publisher
Jennifer Codner, Senior Field Acquisitions Editor
Michelle Piehl, Senior Project Editor
Abbey Hastings, Associate Production Editor
Emely Villavicencio, Senior Graphic Designer
Stephanie Kohl, Licensing Coordinator
Gustavo Youngberg, Interior Designer
Natalie Piccotti, Director of Marketing
Kassie Graves, Vice President of Editorial
Jamie Giganti, Director of Academic Publishing

Copyright © 2019 by Cognella, Inc. All rights reserved. No part of this publication may be reprinted, reproduced, transmitted, or utilized in any form or by any electronic, mechanical, or other means, now known or hereafter invented, including photocopying, microfilming, and recording, or in any information retrieval system without the written permission of Cognella, Inc. For inquiries regarding permissions, translations, foreign rights, audio rights, and any other forms of reproduction, please contact the Cognella Licensing Department at rights@cognella.com.

Trademark Notice: Product or corporate names may be trademarks or registered trademarks, and are used only for identification and explanation without intent to infringe.

Cover: Copyright © 2018 iStockphoto LP/imaginima.
         Copyright © 2018 iStockphoto LP/ktsimage.
         Cover: Copyright © 2018 iStockphoto LP/metamorworks.
Back Cover: Copyright © by Jessie Holloway Photography. Reprinted with permission.

Printed in the United States of America.

ISBN: 978-1-5165-2743-4 (pbk) / 978-1-5165-2744-1 (br) / 978-1-5165-7294-6 (HC)

Brief Contents

1   Why Are We Writing This and How to Use This Book   1
       by Ashleigh & Geoff

2   What You Need to Do as an Undergrad to Get Into Grad School   5
       by Ashleigh
       Ashleigh's interview with James Brockmole, PhD

3   How to Interview With and Select a Graduate Mentor   37
       by Geoff
       Ashleigh's interview with Julie Golomb, PhD

4   How to Succeed in Grad School   57
       by Geoff
       Ashleigh's Interview with Yuhong Jiang, PhD

5   How to Use a Postdoc as a Launching Pad   81
       by Geoff
       Ashleigh's Interview with Elizabeth Buffalo, PhD

6   How to Get a Job at a Teaching School   97
       by Ashleigh
       Ashleigh's Interview with Lauren Hecht, PhD

7   How to Get a Job at a Major Research Institution   125
       by Geoff
       Ashleigh's interview with Lera Boroditsky, PhD

8   How to Succeed in Your New Teaching Job and Get Tenure   149
       by Ashleigh
       Ashleigh's Interview with Glenn R. Sharfman, PhD

| 9 | How to Succeed in Your New Faculty Job and Get Tenure at a Research Institution | 175 |

by Geoff

Ashleigh's Interview with Timothy P. McNamara, PhD

| 10 | How to Get a Grant | 199 |

by Geoff & Ashleigh

Ashleigh's Interview with Mark D'Esposito, MD

| 11 | How to Run a Lab Without Graduate Students | 221 |

by Ashleigh

Ashleigh's Interview with Jeremy Wolfe, PhD

| 12 | How to Have a Great Life and a Great Career: The Teeter-Totter of Work-Life Balance | 237 |

by Ashleigh

Ashleigh's Interview with Caitlin (Hilliard) Hilverman, PhD

Glossary	259
Acknowledgments	263
The Conversation Continues on Brain Bios Podcast	265
Index	267

Detailed Contents

1   Why We Are Writing This and How to Use This Book   1
  by Ashleigh & Geoff

2   What You Need to Do as an Undergrad to Get Into Grad School   5
  by Ashleigh
  - The differences between master's and doctoral programs   5
  - Doctor of Psychology   7
  - The importance of having at least one shiny number   9
  - GRE prep   11
  - Double majors, minors, and extracurricular activities   11
  - Get involved in research!   13
  - Why you probably like research more than you think   13
  - The best-case scenario   15
  - Funding your research experience   16
  - Paid summer research programs   17
  - How to spin most internship and research experiences positively   18
  - How personal to get in your personal statement   18
  - Contact potential faculty mentors in advance   19
  - Psychology vs. neuroscience   22
  - How to ask for letters of recommendation   23
  - Holiday and birthday presents   25
  - Application Fees   25
  - Is my record strong?   26
  - The wide-open back door to graduate school: the paid research assistantship   26
  - Stay in touch with your letter writers   28
  - In closing   28

  Ashleigh's interview with James Brockmole, PhD   29
  - On GPAs and undergraduate course selection   29
  - GRE   31
  - Undergraduate research experiences   31
  - Funding your undergraduate research experience   33
  - Contacting potential graduate mentors in advance   33
  - Speaking with current graduate students   34
  - Selecting your graduate program   34
  - Application and Interview Mistakes to Avoid   34

3 How to Interview With and Select a Graduate Mentor   37
by Geoff

- The mentorship model   37
- Becoming an independent scientist   38
- Before the interview   38
  - Determining the best fit   39
  - Publication record   40
  - Grant funding   40
  - Personal reputation   41
  - Interview timing and travel   41
  - Preparing for meetings   42
  - What do I wear to interviews?   44
  - The schedule   44
  - Mentoring styles   45
  - Misrepresenting your qualifications   46
- Post-interview   47
  - The offer process   47
  - Making your final decision   48
  - Declining an offer   51
- Enjoy the ride   52
- *Ashleigh's interview with Julie Golomb, PhD*   52
  - On the issue of fit   52
  - Why interviews matter   53
  - Selecting a junior versus a senior mentor   54
  - Questions to ask your potential mentor   55
  - Things your potential mentor wishes you knew   56

4 How to Succeed in Grad School   57
by Geoff

- What is graduate school?   57
- How success is measured   58
- Getting your feet underneath you   59
- When should you start acting on your own ideas?   60
- Don't collect data without having your analyses worked out   61
- Why having multiple projects going at the same time is essential   63
- Why you should volunteer to give lots of talks in grad school   65
- Rules of thumb on giving talks   66
- Becoming strict with your time management   67
- How to deal with hard times or quit graduate school   69
- Things to do now to develop your career for the next stage   71

Ashleigh's Interview with Yuhong Jiang, PhD   74
- Formative graduate school experience   75
- Most important quality   75
- Common stumbling blocks   75
- Dealing with student-advisor conflict   76
- Distinguishing between undergrad and grad school   77
- Time management   77
- Grant writing   78
- Preparing for life after graduate school   79

5   How to Use a Postdoc as a Launching Pad   81
by Geoff
- What is a postdoc?   81
- Postdocs increase competitiveness for faculty jobs   81
- Finding a postdoc   83
- What will you learn, Dorothy?   84
- Skills for the future   85
  - Writing   85
  - Learning the methods   86
  - Getting grants   86
  - Mentoring style   87
- Have an explicit conversation with your mentor about independence   88
- Have that grant ready   89
- Defer your start date to the following year   90
- Summary   90

Ashleigh's Interview with Elizabeth Buffalo, PhD   91
- Beth's postdoctoral experience at NIH   91
- What postdocs bring to the table   92
- What Beth is looking for in a postdoc   93
- Opportunities for trainees   94
- The benefits of having a postdoc in the lab   94
- Expectations of independence   95
- Tips from Beth   96

6   How to Get a Job at a Teaching School   97
by Ashleigh
- What is a teaching job?   97
- How to prepare for a teaching job   100
  - Getting teaching experience   100
  - Mentoring students in the lab   102
  - Teaching postdoc   102

Preparing your application *103*
- *Institution-specific research plans* *103*
- *Cover Letter* *104*
- *Teaching Statement* *104*
- *Diversity Statement* *107*
- *Research Statement* *108*

Interviewing *109*
- *Classroom range* *111*
- *Negotiation* *112*

Is a nontenure-track job so bad? *113*

Conclusion *114*

Ashleigh's Interview with Lauren Hecht, PhD *115*
- Suggestions on getting teaching experience *115*
- Cover letter *115*
- Questions you will be asked in your interview *116*
- Questions you should ask during your interview *119*
- What you can be doing during graduate school *120*
- Campus Interviews *121*
- Kiss of death *122*
- Negotiation *124*

7   How to Get a Job at a Major Research Institution   *125*
by Geoff

- How to get an interview *126*
- The ads *127*
- Parts of the application *128*
- The job talk *130*
- The interview *131*
- The chalk talk *134*
- Negotiating the offer *134*
- You've accepted an offer, now what? *137*

Ashleigh's interview with Lera Boroditsky, PhD *138*
- On taking the plunge *138*
- Interviewing the institution *139*
- Two-body problem (and opportunity) *139*
- Mentoring statement *141*
- Teaching *142*
- Responding to questions during the job talk *142*
- Lera's other job talk tips *143*
- Be ready to engage everyone *143*
- Framing concerns *145*

    Confidence *145*
    Negotiation *146*
    Conclusion *147*

8 How to Succeed in Your New Teaching Job and Get Tenure 149
   by Ashleigh
    The candy bowl *150*
    Teaching evaluations *151*
    Time management *154*
    Your research protects and promotes you *155*
     Authorship issues *156*
    Establishing boundaries *158*
    Tenure expectations *159*
     Tenure interview questions *159*
   Ashleigh's Interview with Glenn R. Sharfman, PhD *160*
    Glenn's take on the candy bowl *161*
    Teaching *162*
    Service *164*
    Research expectations *165*
    How much is enough? *167*
    Unique issues while getting tenure at a SLAC *168*
    Rigor and value added *171*
    Administrator's issues *172*

9 How to Succeed in Your New Faculty Job and Get Tenure at a Research Institution 175
   by Geoff
    Research above all *176*
    Lay out your priorities a grant proposal *177*
    Protecting and managing your time *178*
    What we didn't teach you *180*
    Advising trainees and setting expectations *182*
    Your trainees hold the key to their success and demonstrate the limits of your influence *184*
    Collaborations and the diversity of your portfolio *186*
    Why you should avoid spending all your time on your classes *188*
    Stay the course *190*
   Ashleigh's Interview with Timothy P. McNamara, PhD *190*
    Do the homework *191*
    First pre-tenure review *192*

- Second pre-tenure review   *192*
- Teaching expectations at an R1   *193*
- What else junior faculty can do besides publish   *194*
- Why would the department vote yes but the dean vote no?   *196*
- Disillusionment   *196*
- Promotion of women and other minorities   *197*
- The goal isn't actually tenure   *198*

10 How to Get a Grant   *199*
by Geoff & Ashleigh

- To whom would you want to give a grant?   *200*
- How to write the proposal   *202*
- Specific Aims   *204*
- Encouragement for the junior faculty   *206*
- Resubmission   *206*
- Grants for primarily undergraduate-serving institutions   *208*
- Talk to the program officer   *209*
- The take away   *210*

Ashleigh's Interview with Mark D'Esposito, MD   210

- Pilot data   *211*
- Alternate outcomes   *211*
- Figures   *211*
- Specific aims figures   *212*
- What is out of your control?   *212*
- What is in your control?   *213*
- First person versus third person and voice of the narrative   *214*
- What to do while the grant is under review   *214*
- Future directions   *215*
- How to establish yourself as an expert   *216*
- Specific Aims   *216*
- How to be repetitive in a useful way   *217*
- Soliciting a co-PI   *218*
- What to do with a revision   *219*
- Serving on a study section   *219*
- Differences between NSF and NIH   *219*

11 How to Run a Lab Without Graduate Students   *221*
by Ashleigh

- Give a talk in the psychology department   *222*
- Create a partnership with a local SLAC   *222*
- Contact master's programs that advertise research experience   *222*

   Apply for summer research program grants *223*
   Network at conferences *223*
   Create a contract with collaborators *224*
   Prep during summer or winter breaks *224*
   Fill your lab with all the other ranks *225*
   Lab memory *225*
   Running a lab at a liberal arts school *226*
   The benefit of running a lab with undergrads rather than grad students *227*
   Why you should bother *229*
  Ashleigh's interview with *229*
   Three levels of personnel *230*
   Full-time paid research assistants *230*
   Postdocs *231*
   Visiting guests *231*
   Working with undergrads *232*
   Successful RA applicants *232*
   Benefits to his system *234*
   Unique summer programs and high school students *234*
   Overall impression *235*

**12 How to Have a Great Life and a Great Career:
The Teeter-Totter of Work-Life Balance 237**
  by Ashleigh
   The pendulum of being present *237*
   Your family can be your '79 Volkswagen Bug or your cactus *238*
   Maximizing your daily rhythm *238*
   Be a grownup about your lists *239*
   Know your best thinking time *240*
   Therapy as a time management tool *240*
   When have you put in enough time? *240*
   Don't let your email own you *241*
   Meetings *243*
   Dealing with infrequent feedback *243*
   When is a good time to have kids? *243*
    Maternity leave and the tenure clock *244*
   Single parents *245*
   Salary *246*
    Serving on a peer-reviewed journal's editorial board *246*
    Write a book or textbook *247*
    Teaching extra classes *248*
    Summer salary or grant funds *248*

   Study section   248
   Speaker honorarium   249
   Service positions   249
  Just say No   249
  Working from home   250
  How to do a geographically targeted job search   251
  Contributing to a family-friendly department   251
 Ashleigh's Interview with Caitlin (Hilliard) Hilverman, PhD   253
  Boundaries   253
  Morale folder   255
  The good life checklist   256
  Self-care   256
  A better way to organize your to-do list   256

Glossary   259

Acknowledgments   263

The Conversation Continues on Brain Bios Podcast   265

Index   267

CHAPTER 1

Why We Are Writing This and How to Use This Book

by Ashleigh & Geoff

In many ways the model of higher education does not allow for undergraduate students, graduate students, or even faculty to easily access the much-needed advice we provide here. Faculty who teach multiple undergraduate courses and are most likely to spend time advising students often do not have time to also be conducting cutting edge research and mentoring graduate students. Therefore, the research faculty who have the answers about how to get into graduate school are not the faculty to whom the students readily have access. Further down the road, graduate students are usually working with faculty at research-intensive universities who are not familiar with teaching-focused careers and can't help advise them about how to pursue a liberal arts career. This prevents many graduate students from getting the career advice they need and it also prevents faculty members from advising their graduate students with the confidence and success they want to have in placing their graduate students in successful positions. The strength of our collaboration is that we have expertise in both of the primary academic career paths: the research career, and the teaching career. In this book we have combined what we view as the most necessary advice for both these career paths. By writing down in one place the answers to what we both know are frequently asked questions across these two career paths, we aim to even the playing field and support as many budding scientists as possible. Power to the people!

Although we have had quite different career paths from each other, between us we still do not know everything, and we cannot represent the diversity of opinion in the scientific fields of psychology and neuroscience with just two authors. To overcome this limitation, we have enlisted friends and colleagues as contributors to this book. Each chapter ends with an interview Ashleigh conducted with an expert on the topic of that chapter. The interviewees will give you additional information about the topic, tell you if their experience has been different from what we describe, and try to provide additional advice to help you through that period of your journey. We thank them for their time and interest in helping prepare you for your future.

We encourage you to use the topics we raise as jumping off points for starting conversations about your career with peers, mentors, students, and colleagues. We obviously think we are giving you good advice, but we also know there are as many opinions in these fields as there are people. Certainly, advice is autobiographical. It's predictable that some people's experiences will cause them to disagree with our advice. We certainly do not write this book with the intention of defending our advice like a fortress. Use the topics, opinions, and questions raised here as a starting point to solicit opinions from people you respect in the field. As many answers as we aim to provide for you, we also hope to inspire you to ask just as many questions of others in the field.

We hope this book provides you with the tools to make a basic map of where you want to go and how to get there. The issues covered in the chapters here have been the topics most needed by our trainees along the modal path from being an undergraduate student to a securing a faculty job. For alternative viewpoints and other topics that we do not cover here, there are excellent resources, such as *The Compleat Academic: A Career Guide*, articles published in *The Chronicle of Higher Education,* and *Science Magazine's* Career News section.

You might look at the table of contents and think that we really should have included a chapter about getting jobs in industry. We agree that this is a very useful topic for any graduate student, **postdoc**[1], or faculty member who does not feel excited or comfortable in our fields of psychology and neuroscience. We talked to enough people with such jobs while preparing to write this book to know that we were not the right people to write such a chapter, or better yet, an entire book itself. What we learned during these discussions with people who work in industry is that there is not a single prescription for getting jobs after you graduate or leave a career in academia. In fact, one of our good friends with an amazing job in the private sector finds the common unitary usage of the term *industry* reveals dangerous naiveté. Jobs outside academia are not some monolithic entity in which all corporations are looking for the same set of skills or characteristics or knowledge. However, all of them are looking for someone who really wants to work on the company's projects. If you interviewed an applicant who said that they couldn't make it in their profession, so they were looking for something easier like what you do, then you would not give them the job. As academics, we would be well suited to stop thinking of nonacademic jobs as some consolation prize for the losers. The best of these jobs are as hard to get as the best jobs in academia. We hope to address this need in the future, but for now, this book provides more options than you may think.

A common perception among those at PhD granting universities and colleges is that there are very few jobs in academia. The great news is that the academic options are much broader than you might realize. In the United States, there are 115 schools that are classified as being in the top tier due to granting PhDs and having the highest levels of research activity. It is true that while this number is large relative to any other country on earth, those universities can provide only a limited number of faculty jobs. However,

---

1 Darley, J. M., Zanna, M. P., & Roediger, H. L. III (Eds.). (2004). The Compleat Academic: A Career Guide. (2nd ed.). Washington, DC, US: American Psychological Association.

there are 334 PhD granting universities and colleges in the United States that have at least moderate levels of research activity. But wait it gets better. Academia is much bigger than just schools with PhD students. There are over 500 small liberal arts colleges that specialize in undergraduate education in the United States alone, more if we also count those in Canada. Imagine training automotive engineers and bemoaning the fact that there were only 900 car companies in the United States with which to find a job. Sound crazy? One of the primary reasons we wrote this book was so that those at PhD granting institutions could learn some insider secrets for how to get a job at one of the 500–600 small liberal arts schools. We hope that this will help keep those who really love academia in the field by opening eyes to often forgotten possibilities.

Finally, we want to make it clear that few of us who are satisfied and fulfilled in our careers could have predicted or planned for every twist and turn that landed us where we are today. This book outlines a career map, but you create your own path. Good luck!

CHAPTER 2

What You Need to Do as an Undergrad to Get Into Grad School

by Ashleigh

This chapter describes what you should do through all four years of your undergraduate career to be competitive for top PhD (doctor of philosophy) programs in your field. This advice follows you through your undergraduate program and up until you submit your application materials for graduate school. Chapter 3 guides you through the interview process for the PhD program and meshes with this chapter on several topics, so it is useful to read them together. One of the purposes of this book is to help you make better choices about how you spend your time now, by helping you to understand what will be needed in the future. For this reason, it is a good idea to read all the way through the book now rather than later, including the material about getting a job after your PhD is complete.

If you are graduating from high school and embarking on your undergraduate career, it bodes well that you are already preparing for the next stage of education. It will be easy to fit in all our suggestions during your undergraduate career. For more advanced students, I realize that you may pick up this book during your junior or senior year, but much of the advice still applies—just make sure you get moving as quickly as possible. Finally, if you are a faculty member working with undergraduate students, this chapter will give you ideas about how to advise students who want to get into graduate programs (and if you are extremely busy, just hand them this book—we will take good care of them).

THE DIFFERENCES BETWEEN MASTER'S AND DOCTORAL PROGRAMS

My years working with undergraduate students taught me that there are a few major categories of graduate programs that undergraduate majors in psychology and neuroscience typically consider. Broadly, the categories of degrees pursued are master's programs, PsyD programs, PhD programs, and medical school. I will not address the path to getting into medical school here since this is not a pre-med book, but I will describe the relationship between these other types of degrees to clarify some common confusion. Since this book is about becoming a scientist, the only degree that is a research degree preparing you to be a scientist is the PhD. Therefore, here

I discuss why neither a master's degree nor PsyD is a sufficient degree for becoming a scientist. However, both these degrees have a purpose and may be ideal for some students.

Generally speaking, the PhD is a gold standard degree in psychology and neuroscience. If you go to websites, you will note that most of the top 150 schools offer a PhD program that you can apply to, and yet they grant master's degrees as well. What this usually means is that they really only have a PhD program. They typically grant master's degrees to the people who leave the program after a couple of years or for a variety of other reasons. Degrees that are granted to students leaving the program are known as terminal master's degrees.

Some schools do offer a separate master's program or track that you can apply to. This is becoming increasingly popular, in part because it serves as a revenue stream for some institutions. That is, the university administrators may strongly encourage departments to create master's programs because they charge the students tuition. In contrast, PhD programs pay your tuition for you, and pay you a stipend (i.e., a small salary) just for attending. In the United States, these stipends vary from about $17,000 to $32,000 USD per year. This should give you an idea about how selective these programs are. Imagine an undergraduate program with no tuition that paid you to study there where the student-to-instructor ratio was 1:1. It would be competitive to get in, right?

There are certainly exceptions to this characterization of master's programs. For example, some master's programs, such as one at Villanova University, consistently offer tuition remission (i.e., discounted or completely waived tuition) and even stipends for enough students that they may be revenue neutral programs. These programs that offer excellent funding packages can be great choices for students who know they want to go to graduate school but are not yet ready to commit to a five-year PhD program. Master's programs may be a good choice for students who need to strengthen some aspect of their PhD application, such as increasing their graduate record examination (GRE) scores or being able to report a grade point average (GPA) from a graduate program because they did not get stellar grades as an undergrad. Students who have not yet decided on a specific area of study in graduate school may also benefit from the advanced course work of a master's program where the depth and breadth of topic coverage may help them select a career path.

Master's programs are generally two-year programs that may be in a specified subfield field (e.g., counseling, social work, marriage and family therapy). If you earn a master's in social work for example, you may be qualified to work as a social worker in a variety of settings like schools, hospitals, nonprofits, or end-of-life care settings. Some master's programs grant a general master's in psychology at the end of two years and these programs often advertise themselves as steppingstones into PhD programs. These are fine options for students who are very motivated to get into PhD programs but don't get into the PhD program of their choice on the first round.

However, I do caution you that if your goal is to get into a PhD program and you end up in a two-year master's program, you will really only be buying yourself one year of graduate school before you are applying again for PhD programs (because you will be applying in the fall of your second year of the master's program). This is tough because the single most important determinant of success or failure in a PhD program is research productivity. It's unlikely that you did not get into a PhD program because you were lacking coursework, but rather because you were lacking research experience. Therefore, in your master's program it is vital that you hit the ground running on your research and publish your research as soon as possible. This can be difficult because you may still be in undergraduate mode, where coursework is the priority. If you spend too much time on your classes in a master's program, the year will quickly pass, and you may not have gotten much, if any, research done.

Another concern with the master's route is that, if you do not get into graduate school after your two years of a general master's in psychology program, you may also not be qualified to be a practicing psychologist in a field that interests you. This is because your master's in psychology program may not have been designed to provide you with specific enough training to be a social worker, for example, that a degree such as a master's in social work (MSW) would have provided. Be sure to consider what your backup plan will be if you do not get into a PhD program, and be sure the master's still seems worth it to you.

I strongly encourage you to get the degree for the job you want without taking on more debt than you can reasonably expect to pay off. It is vital that you understand what careers are available with the degree you are pursuing and what they pay. It is unwise to take on a huge amount of student loan debt for a degree that positions you to get a low-paying job.

DOCTOR OF PSYCHOLOGY

PsyD stands for doctor of psychology. Alternatively, the doctor of philosophy (PhD) is the more established advanced degree in our field. The PsyD is essentially a doctorate in psychology for people who do not want to conduct research. As a less-established degree, it does not carry the respect and weight in science that a PhD holds, and is insufficient for a faculty position at most research universities. Although I have had many students express interest in PsyD programs, I have three main reservations about the choice to pursue a PsyD.

First, PsyD programs are expensive, very expensive. These degrees are often granted by for-profit institutions (e.g., the Chicago School of Professional Psychology) rather than by your typical public institution that awards master's and PhD degrees. The huge amount of debt that most students will take on to earn their PsyD may not make sense given their career goals or potential future income. For students who are interested in seeing patients and don't want to conduct research, I think that it often makes more sense to earn a master's degree than a PsyD. I realize there are plenty of exceptions to this (e.g., I have a former student pursuing a PsyD at the University of Indianapolis), but as general advice, I am concerned about taking on the student loans required to earn a PsyD, balanced with the earning potential the student will have following graduate school. As a clinician, earning potential may be comparable for someone with a master's in social work and a PsyD, but the debt will be much higher for the PsyD, so unless there is a compelling reason to do a PsyD (e.g., to be a hospital executive), it's probably better to do the master's.

Second, if students are going to invest the years and money into earning a PsyD, why not get a PhD and save the money as well as have the flexibility to potentially have an academic job one day? As discussed in the next chapter, PhD programs likely pay you a stipend and cover your tuition. They are not just free, you also earn a livable salary when you are in those programs. Then you also benefit from having more flexibility in your career down the road should you decide that, for example, you would like to teach at a public university that requires faculty to hold a PhD. If a PhD program is unattractive due to research, then under most circumstances a master's is advisable.

Third, there is a benefit to being in a clinical PhD program when it comes to successfully graduating from the program. Clinical PhD programs help you secure internships (the final step to graduation), and students in PhD programs are more likely to match with a desirable internship. An internship *match* means that one of the internship sites you pick also picks you according to rank ordering (which you can read about online),[1] and you go work with them for a period of time in order to gain the experience seeing patients before graduation. On average, students in PsyD programs have lower internship match rates (75%) than PhD students (82%).[2] Of the 75% of PsyD students who match, only 59% actually match to an American Psychological Association (APA) accredited internship, compared to 91% of PhD applicants. Internships are required for graduation in most clinical programs. Students who do not successfully match either have to keep paying tuition in their PsyD programs until they do match, or drop out. The match rates cited are averages, so if you are considering specific PsyD programs, be sure to check how successfully the programs you're considering match their students to internships.

As I said earlier, unlike a PhD, earning a PsyD will limit potential job prospects in a way that doesn't make sense to me—someday you might decide that having a job with summers off sounds like a pretty nice gig (i.e., as a professor). On the other hand, to be fair to PsyD programs, advantages of a PsyD program may include (1) a focus on developing your identity as a clinician, (2) opportunities to train and practice in a variety of

---

1 https://www.appic.org/Match/FAQs/Intro-to-the-Match
2 www.appic.org, 2011–2016 match rate data

settings with various populations before internship, (3) being taught by professors who are also actively seeing patients and bring that experience into the classroom, (4) considerable classroom discussion about tying theory to cases, (5) some programs require dissertations, so students may still learn to critically evaluate and conduct research, and (6) actively working with clients early on in graduate school, leading to a high degree of experience and comfort upon graduation. Although these are potential strengths of PsyD programs, these opportunities are also be available in some PhD programs, such as the dissertation, so do your homework on specific schools.

So now that you understand the topography of the landscape a bit, how do you position yourself to get into these programs? From here on out I will focus on advice for getting into PhD programs. I do this because this advice generally translates to PsyD and master's programs, which often look for similar qualifications, but are less selective. In other words, if you follow my advice and you prepare yourself to get into a top PhD program, you will also have very good odds of getting into a master's or a PsyD program.

THE IMPORTANCE OF HAVING AT LEAST ONE SHINY NUMBER

I advise my undergrads that they need at least one shiny number in order to get a look from potential mentors reviewing applications in the top PhD programs. That number refers to your GPA or your GRE score. There are many books that prepare you for the GRE, so I will not reinvent the wheel and give you much advice on that exam here. In addition, there is something of a trend for programs to weight GRE scores less heavily than GPA and other factors in their assessments of applicants. That does not mean that you can skimp on preparation. However, it means that GPA, research experience, and publications are even more important than the GRE. Next, I want to emphasize why you need to protect your GPA and how to do so.

Protecting your GPA means not engaging in activities that lower your GPA if they don't benefit your graduate school applications. Very few programs are impressed with your 18 credit-hour semester, so do not burden yourself with that kind of load. When you have to take a particularly challenging course, make sure you balance it out with some courses that will require significantly less attention. Summer session is a great time to take courses that need your full attention to get an A. I know your parents and your wallet might be impressed if you graduate a semester early, but graduate school mentors are not impressed, especially not when graduating early comes at the cost of overloading your semesters, tanking your GPA, and having less time for important research experiences, as discussed later in this chapter.

While I do want to encourage you to embrace all the learning opportunities that appeal to you in college, I also want to encourage you to take advantage of auditing courses or embarking on learning opportunities that will not wreak havoc on your GPA. Your advisor is your friend here and will be able to inform you what your specific institution offers so you can take something outside your required area without making yourself vulnerable to lowering your GPA. Remember that these unusual courses you select out of pure interest are for you, and they are unlikely to impress a graduate program. Nobody cares that you branched out to take German Level 5 if it ended up being the only grade in your four years of college that was not an A (in our house this is known as pulling-the-Ashleigh).

It's definitely worth your time to become a well-rounded individual and a unique applicant. Learning a foreign language or studying abroad, for example, are wonderful opportunities you can and should embrace while in college—I am simply suggesting that you carefully balance these opportunities with your graduate school goals. My only regret from my undergraduate days was not studying abroad, so I wholeheartedly encourage you to engage in those opportunities. But I suggest you try to do it earlier in your education, ideally as a sophomore or during an abbreviated semester offered at your institutions, such as Janterm (a condensed course between fall and spring semesters), summer session (summer school), or Maymester (a condensed course after spring semester but before summer school begins). This will allow you to have the experience, yet not have it interfere with working in research labs as a junior and senior. It is that work in research labs that will be the most important aspect of your graduate school application, as discussed in a subsequent section.

Students ask me all the time to give them an example of a solid GPA or GRE score, but the answer is that it depends on where you are applying. Most graduate programs will list the average GPA and GRE score on their website for their most recent incoming class or during the last several years. So, if you want to know what your scores should be, or whether a program is a good fit, look specific programs up online and gauge whether your scores are consistent with their most recent cohort of accepted students. If they do not list scores online, you can email the contact person listed for the program. However, do your homework first. When the Director of Graduate Studies knows that the answer you seek is on their website, sending an email to tell you the answer is not a particularly high priority for them and quite frankly, indicates that you're unable to research information online for yourself.

Another frequently asked question is whether GPA scores are weighed more heavily than GRE scores or vice versa. Although I can't speak for everyone, faculty mentors do realize that your GPA is a cumulative index of your work across your entire undergraduate career and your GRE is a standardized test score from one day. If your GPA is very high, that's an impressive statement of your stamina, ability to study and focus, as well as your intelligence. If your GPA is low due to one rough semester or starting in a different major, it might help if you work this explanation into your personal statement and ask your letter of recommendation writers to do the same.

People also recognize that GRE scores do have systematic biases. That is, people from underrepresented minority groups in the United States have been shown to have lower scores on average[3]. This happens in virtually all standardized intelligence tests, of which the GRE is really just a special one. However, GRE scores are still frequently used as an index of how quickly and carefully a student can learn a bunch of seemingly irrelevant material. The idea is that if they can learn quickly, they will be good at picking up the information in the lab too. Although there is a weak correlation between GRE scores and ultimate success in graduate school, do your best to keep both numbers as high as possible.

GRE PREP

I told you above that there are loads of books and websites that help you prepare for the GRE (I have some students who swear by magoosh.com), so I will not discuss at length how to prepare for this important exam. I just want to give you a couple of specific suggestions for studying. First, study as early and often as possible. I would get started on this a year in advance of the time you plan to take the GRE. Many students decide to take the GRE for the first time over the summer before their senior year, so that means the summer before your junior year is a good time to start studying.

Secondly, there are many ways you can improve your GRE score over time with study habits that you build into your day. Increase your vocabulary by using a word-a-day app and challenge yourself to text your mom, dad, or friend every day, correctly using the new vocabulary word. Further, take advantage of the *test effect*. We know that the best way to perform well on tests is to take practice tests. It turns out that taking tests is actually better than restudying the material. This is a counterintuitive finding because when you restudy you have access to all the correct information, but when you take a practice test, it feels like you might forget the correct material. However, testing improves memory for the targeted material better than rereading it, provided that at least two days pass between when the material is learned and when testing occurs.[4] This two day time lag is always the case for the huge volume of information you need to learn for the GRE. Therefore, the best reason to purchase GRE prep materials is to get access to the practice tests. Test yourself early and often.

DOUBLE MAJORS, MINORS, AND EXTRACURRICULAR ACTIVITIES

Undergraduates often get excited about the possibility of double majoring or picking up a minor. Ask anyone you know who has been out of college for a while how many times they've been asked what their major was in school—the answer will be basically zero.

---

3 This is a well-documented effect that is not due to intelligence but due to cultural, environmental, structural and governmental factors, to name a few.
4 Roediger, H. L., & Karpicke, J. D. (2006). Test-enhanced learning: Taking memory tests improves long-term retention. *Psychological Science, 17*, 249–255.

There is an Oprah Winfrey clip where she confirmed that, in her entire wildly successful career, nobody ever asked her what she majored in during college. I know you're probably wondering why I'm talking about Oprah when you're trying to get a PhD. My point is that while you are in college, it's hard to realize how little the world is going to care about the details of your majors and minors once you graduate.

People looking at your graduate school application are unlikely to be impressed by your double major or your minor. Although admissions committees for master's, PsyD, or even some PhD programs that do not offer rigorous research training might be swayed by a double major, remember that in order to be a scientist, research matters above specific coursework. However, it is very advantageous to take classes outside your major (or audit them if they might lower your GPA) that you've been advised will be useful to you in graduate school. Specifically, in many psychology and neuroscience programs, it's imperative to know how to write code, so a rigorous computer science course (not just one that teaches Excel) would make you stand out among the rest of the applicant pool. In other disciplines, it might be a particular kind of statistical software that is taught in an upper level statistics course. If you are interested in doing computational modeling of the brain, then advanced math classes would be favorably looked upon by potential mentors. In order to get publications, you also have to know how to write scientifically. Computer science, math, statistics, and English are probably the only minors that you should look at as being potentially valuable for graduate school. It may be tempting to think that you only need four more classes to get a minor in music, but unless you plan on working in a lab that studies the cognitive processing of music, then it is probably an additional course load that could hurt you more than help you.

You want to become a unique package because it sets you apart from other applicants. I had a potential PhD advisor get excited about my German speaking ability because he had collaborators in Germany. Just make sure that your unique package does not come at the cost of your GPA. If those music courses I mentioned above are a breeze for you, then they may well help you to increase your GPA with some extra A's and build that unique angle to help you get into labs that study music perception and cognition, so go for it.

Do extracurricular activities help your application? Generally, not. I know many sources talk about the importance of extracurricular activities to get into the top undergraduate institutions, and they are important for med school, but it's not the same for graduate school in psychology and neuroscience. A few extracurricular activities that make you an interesting, well-rounded person who can say interesting things during an interview weekend are awesome. But small talk is about all they buy you. Indeed, some people believe that extracurricular activities could be evidence that a candidate lacks focus and will not succeed in graduate school. The logic of that point of view is that extraneous interests will keep you from focusing on your graduate work. Because opinions on extracurricular activities can vary, think of extracurricular activities as stuff you do during your free time for exercise or to keep you sane, but they will generally not go into your application. Finally, I think that if you are going to be an active member of a club, get a leadership position. A leadership position might help convey you're willing to work hard, you are organized, you can collaborate with others, and are willing to lead if circumstance necessitates it. A leadership position also gives you something else to mention in your email when you reach out to potential faculty mentors (which I discuss later in this chapter). Don't choose to take on a leadership role over research experience if you have to pick one or the other. It's hard to know if anyone will be impressed by that role, so I'm simply suggesting that if you are going to take the time to attend the meetings and functions of a club, and being on the board doesn't take much more time, then you might as well do it. I was president of the Purdue Equestrian Team and, while I doubt it had an impact on my graduate applications, it was a great outlet for me to get exercise and socialize, which in turn made me a better student. I was also president of Psi Chi, which helped me get to know the faculty advisor well and her personalized advice on my graduate school interviews changed the trajectory of my life for the better. On the other hand, Geoff did not hold any leadership roles during undergrad and he obviously turned out just fine.

GET INVOLVED IN RESEARCH!

If you are reading this book because you want to go to a PhD program, I probably don't have to tell you that a PhD is a research degree. That means that in order to convince potential graduate schools and mentors that you are truly interested in research and know what you are getting into, you need to conduct research as an undergraduate. However, not all research experiences are equal. A research project you conduct in a research methods course, or even in a senior capstone course, is rarely publishable and is not as impressive as working in a lab on publishable research. Similarly, presenting at your institution's undergraduate research conference is not as impressive as presenting off campus at one of your field's major conferences (e.g., the annual meeting of the Psychonomic Society). This is certainly not to discourage you from nailing your research methods project (in fact, you better get that A!) or presenting at your institution's undergraduate research symposium (a great time to try out that new pantsuit you got for Hanukkah) because any research experience is much better than no research experience.

WHY YOU PROBABLY LIKE RESEARCH MORE THAN YOU THINK

While I am on the topic of undergraduate research experiences, let's talk about why your experience conducting research in your undergraduate courses might make you think that you hate research.

An undergraduate-level class that provides an introduction to research projects usually involves a classroom full of students working on projects that they, in some way, created themselves. This means that the faculty member leading the course is probably not an expert on most (or any) of the projects being conducted. This creates an inaccurate impression of what real research is like in a laboratory. Specifically, research conducted in a PhD program will be overseen by a mentor who is an expert in the field. This is particularly true during your first several years in graduate school when you will likely start by working on existing projects, and only later will you be expected to branch out and gain independence with your own ideas. So real research does not involve throwing you into the deep end of the pool with no idea of how to swim (i.e., no idea what makes a research question good). But for some reason this is how many undergraduate research methods classes are taught, including some of those I have taught. These research methods projects can be scary, stressful experiences where you feel lost. You have to get the project done by the end of the semester, so you often finish with an incomplete data set and little idea of how to interpret the statistical tests that a graduate student or software provided you. If this sounds familiar, you are not alone. Research in graduate school will hopefully be nothing like this.

Since I am suggesting that you do not judge research (and your potential love of graduate school) by your undergraduate coursework, what should you use to determine your potential love of research? You can earn credit toward graduation (and sometimes an A toward your GPA) for working in a lab during the semester. Although you will often be interviewed in order to be placed in a lab, you should also interview the graduate student or faculty member with whom you hope to work. Ask specific questions about what you need to do in order to earn co-authorship on the project you will work on and if you will be able to present the research at a conference. Again, any experience is better than none, so if you don't have offers from multiple labs, take what you can get. However, if you can select between different labs to work in, be sure to work in the lab where they are going to help you build your CV (a curriculum vitae, or CV, is an academic resume).

Again, research experience is important, so get any that you can, but if possible, get experience in a lab with someone who is studying a similar topic or employing a similar technique that you might want to use in graduate school. This will help you to convince a graduate program that you know what you are getting into (but see the interview with Dr. Julie Golomb at the end of Chapter 3 for an alternative viewpoint). It also increases the chance that the faculty mentor supervising your undergraduate research would know the faculty mentors you apply to work with in graduate school, which in turn, may increase the chance that their letter of recommendation would carry more weight.

What if your interests change after you've started working in a lab as an undergrad? Faculty mentors are likely to be understanding if you decide to move into a different lab to diversify your experience. As long as you are respectful and keep them in the loop, there should be no hard feelings, so pursue your interests. In a section below, I talk about how to put a positive spin on undergraduate experiences that ended up not being in fields you decide to pursue. First, I want to talk about what we would advise our own children to do if they wanted to get into graduate school in psychology or neuroscience.

THE BEST-CASE SCENARIO

The best thing you can do is to work in the lab of a professor for a couple of years as an undergraduate. This can usually be done for credit hours in your psychology or neuroscience major, as mentioned above. This is typically by enrolling in a class called a research practicum, directed study, independent study, or some similar title. Ideally, you would begin gaining research experience as a sophomore. Why? Because at the end of your sophomore year, you can then apply to do an honors thesis in that lab during the next two years.

Three years of research, you say?! Yes. But this is simply the best-case scenario. Why? Because it allows you to do enough research in a lab, that hopefully is turning out publications, which might translate into an opportunity for you to author or co-author one or more papers. Of the people who started working in Geoff's lab by their sophomore year, all left with at least one publication. These publications are the precious jewels in your graduate school application.

How do you become an author on a paper as an undergraduate? The threshold for authorship varies a bit from lab to lab, but the overarching principle is that to be an author, you need to make an intellectual contribution to the work. This can come in many different forms. Performing at least some of the data analyses may be sufficient. You can make a keen observation that results in the idea for a control experiment or a new experiment for the project. You can program the experiment. Typically, these are the kind of hard-thinking things that result in authorship. Data collection rarely does. So, you can begin to see why it is ideal to be in a lab for a few years if you want to acquire the thinking, programming, or analysis skills to warrant authorship on a scientific publication. Be sure to have these conversations with the professor or graduate students you are working for at the beginning of each semester.

Sometimes your honors thesis project will translate into a publication. However, not all publications are created equal, just like the conferences that I mentioned above. Publishing your honors thesis in an undergraduate journal is good. It shows you can carry a project through to completion. Journals, such as *Psi Chi Journal of Psychological Research*, publish undergraduate research projects that faculty mentors do not intend to take to the next level. And many others are listed on the Council on Undergraduate Research's website under the heading "Undergraduate Journals."[5] Getting your honors thesis published by a higher impact peer-reviewed journal in the scientific field would

---

5   http://www.cur.org/resources/students/undergraduate_journals/

be more prestigious than publishing in an undergrad journal, but not all projects warrant such publication. This can occur if there was a significant error made while collecting data, or you don't have enough data collected by the end of your time in the lab, or the general motivation for the project is simply not very impactful. However, you should tell the professor running the lab that your goal is to author a paper so they can help steer you to a project that has a good chance of being published, and help you learn the skills that will allow you to cross that threshold for intellectual contribution that warrants authorship.

If your time in a lab does not translate into publications or an honors thesis, do not panic. The year or two that you spend doing research will give you plenty to talk about at interviews. They may have resulted in conference presentations, which are also good, though not as great as a publication. And your excellent GPA or shiny GRE score will get you a look. This is good news because sometimes, even with years spent in a lab, projects do not work out and cannot be published. Even the most well-intentioned mentor can't promise you a publication by a certain deadline, like by the time you submit your graduate school applications. Remember the quote that people attribute to Einstein, "If we knew what it was we were doing, it would not be called *re*-search, would it?" But sometimes you will work hard and get lucky, and a publication in a top scientific journal can be the thing that gets you interviews at a number of top schools.

FUNDING YOUR RESEARCH EXPERIENCE

You are never too young to fund your own research! My undergraduate students have been applying for and receiving Psi Chi grants for years. Psi Chi (psichi.org) is a wonderful honors society in psychology. You are likely eligible for Psi Chi, since many of you are intending to go to graduate school in psychology. You absolutely must join Psi Chi and apply for grants if you are eligible. If you get a grant, your one-time membership fee will have paid for itself. There are grants that fund research projects, both at your home institution and for you to travel to another institution to conduct research. There are grants that fund travel to regional APA conferences. There are even grants for getting software to run your experiment. Funding your own research project or summer opportunity looks very good on graduate school applications and provides great material for your letters of recommendation.

Another great resource for funding your research would be your institution's undergraduate research office, if it has one. Such offices usually host open houses that inform you about funding opportunities, undergraduate research conferences where you can learn about funding for your research, email lists that notify you of awards, and potential paid research opportunities in labs throughout campus.

Finally, there are a number of great funding resources that are either explicitly for minority students or give preference to minority candidates. Minority status can include individuals from underrepresented groups such as families of low socioeconomic status and individuals with physical disabilities, in addition to the underrepresented minority populations typically considered (e.g., African Americans, Hispanics, or women). For example, the McNair Scholars Program, funded through the United States Department of Education, has approximately 150 institutions across the country.[6]

PAID SUMMER RESEARCH PROGRAMS

I know that it is hard to further your academic and career goals over the summer while also needing to earn money. It can be particularly difficult to get paid for internship experiences in the field of psychology. There are, however, a number of paid opportunities to conduct research over the summer. You just need to know where to look. One popular opportunity is the **National Science Foundation (NSF)**[7] program called "Research Experience for Undergraduates" (REU).[8] Another paid summer research opportunity is the Summer Undergraduate Psychology Research Experience Grants through the APA.[9] The NSF and APA opportunities are both available at multiple institutions and are a great opportunity to travel to other institutions and network with a faculty member who can later write a letter of recommendation on your behalf.

Your home institution might offer funded undergraduate programs as well. For example, Ohio State University (OSU) offers a summer undergraduate research program for underrepresented minorities[10] as well as other paid summer opportunities that can be found online through the OSU Office of Undergraduate Research and Creative Inquiry. You will want to be sure you understand what the money funds. It might pay you for conducting research hourly, it might pay for your room and board, or it might pay for research expenses (e.g., subject money, traveling to conferences, a laptop computer). If you can't cover your own housing, for example, you need to be sure that the research opportunity includes housing. If you can't get any of these opportunities, it would be wise to talk to your advisor about how to spend your summers so that you are still preparing for graduate school. I usually have a few students volunteering in my lab over the summer who either live with their family in town or are working nights and weekends to pay room and board. The advice on how to spend your summer might differ depending on your specific career goals, so here I'll say that at the very least, buy a GRE prep book that offers sample questions and work through the practice tests.

---

6   for more information go to https://mcnairscholars.com
7   Check the glossary in the back of this book for full definitions of terms in bold.
8   https://www.nsf.gov/crssprgm/reu/
9   http://www.apa.org/about/awards/undergraduate-research-opportunity.aspx
10   https://gradsch.osu.edu/outcomes-innovations/summer-research-opportunities-program

HOW TO SPIN MOST INTERNSHIP AND RESEARCH EXPERIENCES POSITIVELY

When I was an undergraduate at Purdue, I interned at a program called the F.O.C.U.S. program (Families of Children Under Supervision)—a program for juvenile delinquents and their families. It was a huge time commitment; each week I spent one night conducting group therapy with the juveniles, one night of group therapy with their parents, and a third night of family therapy with the juveniles and their parents. It was an exhausting experience that helped me know I was not cut out to be a clinical or counseling psychologist. Geoff had a similar undergrad experience working for a crisis center in Iowa City. These were typical experiences because many psychology majors think they want go into clinical psychology to help people or study a disease model in neuroscience. This experience was vital for me because it showed me that I felt burned out and disenchanted after a relatively short amount of time in the clinical world. It might seem like I wasted my precious time outside of classes engaged in this experience because I ultimately decided to pursue cognitive psychology PhD programs. That's not entirely true. As I mentioned above, part of getting into graduate school is illustrating to the program and mentor that you know what you are getting into and you really want to do it. Having experiences in other areas of psychology is one way of proving you are passionate about a particular area and that you know you are not passionate for another. These internship experiences outside of my and Geoff's ultimate area of graduate study were great material for our personal statements since we could essentially describe our exposure to several different career paths, were confident in our choices, and were dedicated specifically to cognitive neuroscience for graduate study.

HOW PERSONAL TO GET IN YOUR PERSONAL STATEMENT

I know there are lots of books and websites with advice on how to write a personal statement. You can check online as well as at your institution's career center for sample personal statements. You might also ask some recent graduates from your institution for their personal statement (provided they got into a program you consider attractive). What I want to share here is some specific advice on what I have seen students include in personal statements that I suggest you avoid.

As I indicated before, many folks get interested in psychology because they want to help people, or they want to know what makes people in their lives behave or function in a certain way. Although that's a reasonable motivation to be interested in a general field when you are 18 years old, you are walking a fine line when you assert that as your motivation to attend graduate school. Let's say that you want to get your PhD in neuroscience because your grandmother has Alzheimer's disease and you want to find a cure. This can be a problematic motivation for a few reasons.

First, think of all the people who have worked on finding a cure for Alzheimer's and yet we still don't have one. The idea that <u>you</u> are going to find a cure for Alzheimer's disease represents a misunderstanding of the rate of science. Many of us love research and want to add to the knowledge base, and we all certainly want to see a cure to Alzheimer's disease. However, if your graduate school hopes are hung on accomplishing a goal you are unlikely to reach, potential mentors may be concerned that your motivation is not sustainable. It takes years to move a project from idea, to experiment, to manuscript, to publication, to application in the real world. Your potential advisor will want to make sure you understand this slow pace of science and be confident that you will not burn out easily and quit graduate school.

Second, if you are applying to a neuroscience program where the faculty are experts on memory and attention, but your personal statement says you want to cure multiple sclerosis, your goal is not on target. It's like saying you want to design a safer car so you're going to pursue your PhD in sociology. You fail to demonstrate an understanding of the field in this example because people who are experts on memory and attention are not working on curing multiple sclerosis. This is important because one of the issues faculty discuss after graduate student interview weekend is how well the applicants know the field, as discussed in Chapter 3. You need to demonstrate breadth of knowledge of your field. That includes understanding both what is *and* what is not studied within your chosen area.

Third, if an applicant provides detailed personal information about themselves, or even a close friend or relative, this can be interpreted as evidence of poor boundaries, which will not bode well for the student in graduate school or life after graduate school. A detailed account of an applicant's mental health struggle, or that of their close relative, does not constitute evidence that they have what it takes to succeed in graduate school. It's more important to the graduate program to assess whether the applicant can think scientifically and critically about clinical problems and whether they have a sense for appropriate clinical boundaries. If the applicant uses the personal statement to detail personal information, scientifically-oriented clinical programs are going to see red flags. Although there are some schools that may prefer and even encourage their applicants to disclose their personal experience with mental health problems, it tends to be a sign of a less reputable program.

CONTACT POTENTIAL FACULTY MENTORS IN ADVANCE

Before submitting your application for doctoral programs, you need to find the people you want to work with, send them emails, and see if they intend to take on a student in the coming year. A given mentor will not take a student each year. If their lab is already full, or if there are other people in the department that need students more, then certain people will not be recruiting in certain years. If you apply to work with someone and design a research statement just for that person who is not taking a student, then your odds of getting a look from someone else who is taking a student are drastically reduced. However, the larger mistake is to apply to a program without aiming your application at specific faculty mentors and failing to name specific people in your research statement who you are interested in working with.

Admittance to graduate programs is largely determined by *fit*, meaning that the individual faculty mentor who would be taking you on has space for you in their lab, has funding for you, and is interested in working with you. Graduate training uses the apprenticeship model. This means that you are typically applying to work with a specific person, not to a program. As Geoff discusses in Chapters 3 and 4, this is because classes are not the primary focus of graduate school. Rather than focusing on memorizing content like you probably did (or will do) to succeed as an undergrad, PhD programs require you to acquire the skills needed to do science.

It is a waste of your time to apply to a program and advisor if the advisor is not accepting new students. Worse yet, many faculty members only look at applications from students who explicitly mention them in their application materials. So, if you do not name any specific faculty members you want to work with, your application may not be seriously considered by anyone. The occasional exceptions to this rule are the interdisciplinary neuroscience programs that have a committee selecting students. Such programs sometimes want students who do not yet have a preference for which faculty mentor they want to work with because the program is designed to require rotations through a variety of labs. But even in that case, the directionless students who do not name potential mentors in their research statements are less likely to get an offer than people whom the committee can categorize as likely succeeding with a few specific faculty members.

Contacting potential mentors ahead of time also helps them be primed to remember your name fondly when they look at the list of applicants. You should do this during your junior year as you are putting together the list of schools to which you are planning to apply. When you send these emails, tell them with whom you are working currently and include a CV that lists your grades, GRE scores (if you have them), and any publications or conference presentations. If you do not have a good idea of what your CV should look like, ask for a professor's input or look at some faculty's CVs that should be easily found on their websites. Your CV should not look like a typical resume.

Ideally, the professor whose lab you currently work in will know this potential mentor. This will allow their recommendation to carry more weight than one from someone they do not know or entirely trust. For this reason, you will want to work with your current mentor to establish a list of people they know and universities with graduate programs.

I suggest saying something simple in your email like:

> Dr. Smith,
>
> I am a junior at Manchester University. I have a 3.82 GPA, am president of our Psi Chi chapter, compete on the equestrian team, and am currently working in the Visual Memory Lab of Dr. Evil (see my attached CV). I am interested in your work on recognition-induced forgetting. Are you taking graduate students for the fall of 2018? If so, are there any particular skills that are requirements for you to consider PhD students, such as knowledge of MATLAB?
>
> Thank you for your time,
>
> Anxious Annie

The number of people you contact definitely varies based on your interest and constraints on how many schools you apply to. I applied to eight PhD programs and got into three of them. These days that number seems low but it was simply the number of faculty I found doing research that interested me. I advise my students to apply to at least ten programs (if not more), ensuring that they are applying to one or two schools they are confident they will get in to. If your goal is to apply to at least ten schools, then you should probably find around 20 faculty to email and ask if they are taking students, assuming some might not get back to you, some will be on sabbatical (i.e., a paid leave where faculty are off campus for a semester or year), and some won't be taking students for other reasons.

Rather than pairing students with faculty members, there are some PhD programs that admit the highest quality applicants to the program as a whole. Although this may be common in neuroscience, it is rare in psychology. Therefore, if it occurs in a psychology program, it may be indicative that the program is not designed to provide high-quality research experiences. If you are lacking research experience, the good news is that these programs also would not have as high expectations in terms of research experience from their applicants. This model can be problematic if you are passionate about receiving mentorship in a specific research topic. For example, you could come into a program having an interest in the high suicide rates of veterans and discover that there are no faculty members who specialize in the military population or that the one faculty member who did specialize in this area has left the school. The typical program that requires applicants to apply to work with a specific faculty member can assess your research interests and aptitude in order to purposefully match you with a research mentor so everyone (both the student and the mentor) can be as productive as possible.

In order to determine if the institution you are applying to does not match with specific faculty, consult their website, check with the director of graduate studies within the department, or consult their graduate school. As previously mentioned, if you send an email inquiring about the process of a particular program, be sure you have done your homework and verified that the information you are requesting is not available online. By the same token, contacting specific faculty indicates a level of sophistication in your application, provided you aren't asking questions that could be answered by searching the web. Contacting faculty in advance indicates sophistication because after having several email exchanges with faculty members, you can comment on those conversations in your personal statement (e.g., "To provide some specific examples of my interest in your program, I'd love to work with Dr. Smith and Dr. Jones. I have been in touch with both Dr. Smith and Dr. Jones and their research aligns very well with my previous research experience and interest in continuing to study visual long-term memory."). However, I've also heard from some faculty at less research-intensive PhD programs that they do not want to receive emails from students because they have hundreds of applicants and couldn't possibly respond to them all. This may be the perspective at programs that admit applicants without pairing individual students with specific faculty mentors. The idea being that there is no reason to correspond in advance since a specific student-mentor match is not being made. Although opinions may vary, my gut reaction is that you won't have a great experience at a school where the faculty do not want to respond to your email, so just send the email and see who responds. This topic is discussed in more detail in Chapter 3.

PSYCHOLOGY VS. NEUROSCIENCE

Many of you who read this book may be asking yourself whether you should apply to work with a faculty member through a psychology or interdisciplinary neuroscience program at that school. Many faculty members are affiliated with both programs and take students through both programs. However, those faculty members probably have an opinion about which program they prefer their graduate students to be in. You should ask potential mentors about which they recommend you apply to because it may impact them, as I discuss below.

There are some fairly consistent differences between graduate programs in psychology versus neuroscience. Most neuroscience programs require, or at least heavily recommend, that students do rotations in a couple of labs before picking a single lab to work in. Although these can be good for students who do not really know what kind of neuroscience they want to do (i.e., human cognitive neuroscience, systems neuroscience, or cellular and molecular), they come at a cost. Usually each rotation lasts a semester or two. While this is enough time to learn what happens in each lab and to contribute a bit to a project, those contributions are often insufficient to warrant authorship on the project. This is because the first couple of months are spent just learning the necessary skills to collect data. Meanwhile, people in the same laboratory that are in the psychology program may be writing up publications as first author by the end of their first year.

The course requirements are often heavier in neuroscience programs than psychology programs, because each of the different departments that formed the program want to make sure the students are getting some training in the topics and material that they see as critical. This time spent on coursework is another reason why you might leave such a program with fewer publications. The other reason why some potential mentors may be less enthusiastic about you applying to an interdisciplinary neuroscience program is that compared to psychology programs, admission decisions for neuroscience programs are more often made by committees. Individual faculty members in the neuroscience program usually have less say in who gets admitted (sometimes, no say at all), compared to psychology programs, where the decisions are made almost exclusively by the potential mentors (assuming the mentor does have an offer to make). The last reason why some potential mentors might prefer you to apply to a psychology program, if you can, is that the cost of students is lower for the faculty member, increasing the odds that they could support you if you were accepted.

It is important to note that if your mentor is a member of both programs, the science you would be doing is probably exactly the same. Only the classes will differ at bit, and classes are only taken during the first couple of years in a program. I am not trying to discount the value of neuroscience programs or the value of those degrees. I realize that this might sound like I have a preference, but I really do not. Instead, I am including this because I think that prospective graduate students might benefit from knowing why a potential faculty mentor might be trying to steer them in certain directions. At some institutions, the neuroscience programs are stronger than those in psychology or brain science. Indeed, there are differences of opinion across faculty at the same place about which program they prefer for their students. As a result, I recommend that you ask the potential mentors which program they would prefer you to apply to work with them through.

HOW TO ASK FOR LETTERS OF RECOMMENDATION

Early on in your undergraduate career you should think about who will write your three recommendation letters. You need to perform well in their classes, make sure they know you, ask to work in their labs, et cetera. It is certainly optimal if at least one of the letter writers is known in their field (i.e., they are actively publishing) and their field is the same area you are seeking to pursue in graduate school (e.g., they study how people store visual information in memory and you want to work with people who study visual memory in grad school). This may occur naturally if your graduate school interests grow out of your research experience as an undergraduate working in a lab.

There is some expected etiquette on how to ask for letters of recommendation. You should first ask each letter writer if they are willing to write you a letter well in advance of the due date, ideally three months in advance. In asking them for the letter, you should remind them of how long they have known you, the courses you have taken with them, your grade(s), the program you are applying to and why you think they will be able to write you a strong letter of recommendation (e.g., the faculty member has seen your research skills in action during a senior capstone course and the job for which you are applying is a job for a **research assistant (RA)** position).

One key trait that graduate mentors are looking for is that you can independently solve problems and overcome obstacles. One concern with students who have high grades and top GRE scores is that they don't know how to fail. Science does not always work out as planned and the mentors want to know you will work hard in the face of adversity. If you have a specific story where you dealt with a crisis in the lab or independently solved a major research problem, remind your letter writer of that incident so they can include it in the letter.

If they agree, you should provide them with addressed and stamped envelopes for any letters that need mailing (many schools prefer letters submitted electronically), a list of deadlines, your CV, a specific list of what you would like them to say, such as skills you want them to highlight, all nicely organized in a folder (or pdf file if no envelopes are required). Then ask your letter writers how frequently and how far in advance of the deadlines they want email reminders.

Make sure you avoid these all-too common errors:

- Do not just copy and paste emails to different professors and neglect to change the name of the professor to whom you are addressing the email when you ask a professor to write you a letter of recommendation.
- Don't enter the professor's information into an online application system *before* you ask them if they will write you a letter and wait for them to respond. You may not realize this, but the online application portal usually automatically emails them as soon as you've entered their information.
- Thank them if they say no that they cannot write you a glowing letter—they just did you a favor! They could have said yes and then written a letter about how you were frequently late and disengaged in class.
- Don't forget to sign any forms that need to be signed before you give them to the faculty member. Having to track down a student to complete a form is an unfair burden to put on a faculty member.
- Don't ask to see your letter. If you don't trust that you are getting a glowing letter from the faulty member, don't ask them.
- Don't forget to send them reminders according to the schedule you agreed upon. When students do not send reminders, it does not seem like graduate school is important to them. Faculty members might feel that you are wasting their time and it reduces their ability to write you a glowing letter.
- Don't suffer from senioritis while you are waiting for them to write you a letter! Seniors often seem to fall apart, get lazy, or burn out, just when they are asking for letters of recommendation. Keep going strong so your faculty letter writer has only good things to say about you.

HOLIDAY AND BIRTHDAY PRESENTS

I bet you didn't expect to find advice on what gifts to ask for on your next birthday in this book. I am going to do you the favor of assuming you are a broke college student (by the way, thanks for spending money on buying this book) and give you the advice I wish someone had given me. If you follow my advice in this chapter, there will likely come a day when you need interview attire for graduate programs. Holidays and birthdays are excellent opportunities to ask your loved ones to build your professional wardrobe. Ask for a blouse, a blazer, dress shoes, et cetera. These are also excellent opportunities to ask for money for application fees for future graduate school applications. I hate hearing that my students can't apply to all the graduate programs they want because of financial constraints. Another one of my favorite professionally-driven gift ideas is to ask for help in beginning to build your online presence with a website and domain name. I love Squarespace—you can get a website for about $100 USD per year and a domain name for about one tenth of that depending on where you purchase it. Nothing looks better to a potential mentor than Googling an applicant and finding a professional presence in which their current research is described, along with conference presentations, and publications under preparation. (You should also take those inappropriate Facebook profile pictures down. If you are unsure whether something is inappropriate, it is.)

APPLICATION FEES

As I mentioned, I really hate to see our students limit the number of graduate schools they apply to because of the financial burden of application fees. Application fees are roughly $100 per school, although this can definitely vary so check out the specific institution's website where you are considering applying. You don't want to get to this point and then fail to submit an application because of a relatively small application fee. You must prepare for application fees in advance. Consider setting aside some of the money you earn over the summer before senior year for application fees. Post on your university job board to get some babysitting gigs that will pay for these fees; if you tell the person who hires you what you are working toward, I bet they might pay you a little more generously or offer you some extra hours. Some graduate school programs waive application fees if you apply early, so try to stay on top of your game and be ready to hit the ground running to make those early deadlines. In addition, preselect a list of schools you want to apply to before you take the GRE so that you can have your scores sent directly to the school for free by indicating them at the testing center. It doesn't do you any harm if you send the scores to schools you end up not applying to, so going into the GRE unprepared with some programs in mind to which you are likely to apply is a waste of this free resource.

IS MY RECORD STRONG?

You should talk to the faculty member who is your major advisor, as well as the person who runs the laboratory that you work in, about what they see as your chances for getting into a graduate program. Ask them to be honest with you, and not just optimistic. This is important because you need to know if your record of research experiences, GPA, and GRE scores are borderline. You need to know this primarily because you will need to apply more widely than someone with a bulletproof set of application materials.

If your GPA is closer to 3.5 than 4 (in the 4-point system), your GRE scores are closer to the 75th percentile than the 99th, and your research experience is limited to collecting data for a semester or two, then you are not out of the game, but you are not likely to be highly competitive in the graduate school market. This means that you will probably need to apply to and target faculty members in lower ranked programs. This might help your chances because these programs get fewer applications than the top programs. So how do you know what the top-ranked programs are?

The National Research Council provides the best rankings of available programs. The council has multiple criteria, but generally the *R factor rankings*[11] are what people look at, and these are re-evaluated every five years. These rankings are published online by the *Chronicle of Higher Education,* among other sources. It is wise to find faculty who you are interested in working with that span the R-rankings. This is a wise move for students regardless of how great you think your application will be. Graduate recruiting can be a bit unpredictable and it is good to apply to some schools in the 100–200th ranked range (i.e., below the top 100 ranked programs).

THE WIDE-OPEN BACK DOOR TO GRADUATE SCHOOL: THE PAID RESEARCH ASSISTANTSHIP

If you are a college senior applying to graduate school, then you should look for full-time research assistantships as a backup plan as soon as your applications are submitted. Why? Although this varies by program, many people who get into top PhD programs these days have been a full-time paid research assistant for a year or more between undergrad and grad school. Some people refer to these as postbaccalaureate, or postbac, positions. I will use the term RA as it sounds less pretentious. This is not to say you can't get into graduate school straight out of undergrad. Some people still do go straight from undergrad into a PhD program, but it is less likely in the top programs. Your graduate applications will be more competitive if you spend a year or two doing research. It should come as no surprise that a year or two of working as a researcher,

---

11   National Research Council (NRC) rankings are viewed more or less favorably, depending on the source you consult. We take an agnostic view of NRC rankings, but mention them to inform the reader of available sources. There are currently two main rankings, R-rankings and S-rankings. You can research both these online.

40 hours per week, people usually learn a lot, and that people who are better prepared are more successful.

In an RA position, you get paid while learning necessary research skills to survive in graduate school, such as computer programming and how to analyze data. Some of the most successful people in graduate school will have done several years of paid RA training before getting into the grad program of their choice. These are great jobs because you make a decent salary with benefits, and sometimes the position comes with funds for you to travel to conferences, especially if you take over a project as your own.

These positions can be difficult to find. You may have been thinking that these sound like sweet gigs compared to paying someone else for a master's degree. You are correct, and that can make finding one difficult if you do not start early and beat the bushes. You can ask around in your undergraduate department to see if anyone is hiring or knows of someone who is looking for a paid research assistant to work in their lab. However, you typically need to scour the Internet. Society job sites and even some for-profit websites will advertise some[12,13]. Another option is to look at human resources websites of individual universities.

A more personal approach is to send emails to people who you have identified as possible mentors for your graduate training. If it is the February after you submitted your graduate applications and you have not heard from the potential mentors you applied to work with, then you should send them an email and say that you are also interested in working with them as a RA if they have an opening. Faculty typically know if they will be looking for a research assistant in the near future, even before ads are posted. For example, their current RA may have been applying to grad schools at the same time as you and the faculty member will already be thinking about replacing them. You might be able to find such a match before they even post an advertisement online. The undergraduate research office at my institution sends out emails with these opportunities somewhat often, and I frequently forward them to colleagues at other schools. This means that letting your professors know you are looking helps them keep their eyes open on your behalf.

To be clear, if you have your heart set on getting your PhD, apply to PhD programs as well as some master's programs as a backup plan. Then once all your materials were submitted, try and find a paid RA position. If you don't get into any PhD programs, prioritize the paid RA position over the master's programs because it allows you to focus on learning research skills (programming, data analysis, and writing) without the worry of coursework. Last year in one of our PhD graduate programs, only 1/8 of the 1st year grad students came straight from their undergraduate school. I know this sounds daunting, but paid RA positions help make you competitive against other applicants who did not work as RAs.

---

12 such as http://neurojobs.sfn.org/jobs
13 such as http://www.indeed.com

STAY IN TOUCH WITH YOUR LETTER WRITERS

Most of us work with undergraduates and RAs because we find mentoring fulfilling and we grow to care about you. Please stay in touch with us. Let us know when you get into a program, or when you don't get admitted to any programs. Let us know how your career and life are going down the road—we LOVE to hear success stories from you!

Once you get into grad school, you should submit an NSF (National Science Foundation) fellowship grant application during your first or second year. These are prestigious grants that typically pay you a higher stipend than your institution will. For these applications you need a minimum of three recommendation letters, and can have up to five. Because you will have only been at your graduate institution for a few months or maybe a year before submitting your grant application, you will need the letter writers from your undergrad program again. So, after the dust settles and you are in grad school or an RA position, send your letter writers an email telling them about how everything is going and thanking them for their time.

IN CLOSING

In closing, I want to explain that my own path into a PhD program often felt like I was moving blindly, so please don't feel like you have to have it all figured out as an undergrad. When I started as an undergrad, I was a business major because my dad wanted me to be and I had no idea what I wanted to do with my life. Actually, I did know; I wanted to train horses, but that wasn't an option according to my father. I quickly realized I wasn't as buttoned up as my peers in the business classes and switched my major a number of times, entertaining majors in German and English, before I landed on psychology thanks to a wonderfully engaging Intro to Psychology professor. I pursued an interest in both sociology and psychology and ended up in honors research programs for both disciplines. It was through these research experiences that I realized I was much more interested in the research methods of psychology and decided at the very last minute to apply only to psychology PhD programs, and abandoned my still very real interest in promoting the social justice issues for same sex couples that I was studying in sociology.

I went to graduate school because I loved school, I was a good undergrad, and I didn't want to enter the real world. This wasn't great logic because the skill set you need to excel as an undergrad is not the same skillset you need to succeed in graduate school. For example, getting a 4.0 GPA in undergrad may involve a lot of memorization rather than application. However, in order to succeed in graduate school, you'll need to be able to apply knowledge to create your own experiments and ultimately, an entire research program. Dr. Brockmole's interview below makes this important point: that getting a 4.0 during undergrad doesn't actually indicate that you will succeed in a PhD program.

> **Ashleigh's interview with James Brockmole, PhD**

visualcognition.nd.edu

James Brockmole, PhD is Professor of Psychology and the Joseph and Elizabeth Robbie Collegiate Chair, as well as the Associate Dean for the Social Sciences and Research in the College of Arts and Letters at the University of Notre Dame. Jim runs a large and successful lab at Notre Dame, with numerous undergraduate students working on their senior theses who have successfully received funding for their projects and have placed into top graduate programs. Jim's appointment as associate dean contributes to his thoughtful approach to undergraduate education and the ideal applicant, which really shines through in our conversation summarized below.

ON GPAS AND UNDERGRADUATE COURSE SELECTION

Jim takes a more thoughtful approach to reviewing graduate school applicants to his own lab than I give mentors credit for in this chapter. Although he agrees that a high GPA is of critical importance, he also likes to see that the student has reached for a challenge. For example, he is more inclined to accept a student into graduate school who has taken a course that is designed to be hard like biology, chemistry, or calculus. Or, rather than an applicant who only took the required undergraduate statistics class, he'd much prefer to see a student who took a risk and took a graduate-level course, like a graduate-level statistics class. Of course, he is not impressed by a student earning a D in such classes, but if they are getting B's in some of those challenging classes he would prefer that to a student who has never really taken anything that looks hard and maintained a 4.0. He acknowledges that this might make more sense for programs like cognitive psychology or cognitive neuroscience, but based on your area of interest, you can inquire with your professors about which courses would demonstrate this reach for a challenge. Jim is looking for an applicant who doesn't look like they are timid and always taking the safe route because science isn't always about being safe. Rather, Jim wants a student who is willing to reach. He says that it's desirable to have an attitude that it's ok to fail as long as you learn from your mistakes and try again. That doesn't mean it's ok to fail your classes, but that it's good to try new and challenging things.

Jim also suggests that students who are interested in pursuing graduate school make sure they are taking writing intensive courses while an undergraduate. He finds that writing is the hardest thing to teach a graduate student, so if an applicant had English as a second maor, or took a number of extra English courses, he knows they can write. Even though most writing classes are not teaching scientific writing per se, Jim sees that as a relatively easy adjustment to teach. In terms of preparing for graduate school while you are still an undergraduate, Jim points out that poor writing is what slows students down the most throughout graduate school. For example, when it's time to write for comprehensive or qualifying examinations, manuscripts, grants, or the dissertation, the most limiting factor that slows everything down is a student who can't write.

Jim agrees that an 18 credit-hour semester is unwise, so senior year is a great time for students to take a graduate-level statistics course to demonstrate that they are reaching and doing so after their GPA has been reported in graduate school applications. This demonstrates that the student is preparing for school to get harder in graduate school. Another option is taking additional English courses, or writing intensive courses, to make efforts to become better writers in preparation for graduate work. Just be sure to try to balance competing demands as best as you can during undergrad. For example, don't kill yourself by taking cell biology and the graduate statistics class in the same semester. Balance is something that is absolutely needed and the sooner you start to think about the competing demands on your course load, the easier it is. You don't want to throw it all in during the junior year.

Students often inquire about how to deal with a rough semester that contributes to lowering their GPA. Jim advises that students can't control the various ways that people are going to assess you, but to take advantage of programs that ask for both your overall GPA and also your GPA in psychology or your GPA from just your junior and senior year. This allows you to demonstrate your focus on the area and possibly explain a low GPA before switching majors or because of a particularly hectic period in your personal life. For example, if someone has a GPA about 3.25 or higher and seems like a good fit otherwise, he's willing to take a closer look at what's going on with them. Jim explains that the reasons that underlie the difference between a 3.5 and 3.8 are varied and aren't necessarily very meaningful. Despite Jim's willingness to dig into the reasoning behind an imperfect GPA, he does point out that at the best programs, students with lower GPAs are going to compete with other candidates who have very high GPAs and who have also prepared well by taking advanced classes. So, he says that students do have to titrate their expectations about the kind of schools they are going to get into.

In addition to Jim's suggestions about course selection during the undergraduate years, you want to remember that undergrad is your one opportunity to learn whatever you want, so take advantage of that opportunity. If you are passionate about foreign languages, take those classes, but just keep in mind that your second major is not going to get you into graduate school, so don't double major for that reason.

GRE

Jim also agrees that although your GPA is a record of four years and the GRE is only a record of one morning, it is one morning that you can prepare for in advance. Therefore, the success on the GRE does reveal both your ability and the preparation you've put into the exam. If you don't do well because you studied for only three days, nobody is going to forgive that because obviously you don't want to get into graduate school very badly. If you are in the position of explaining a poor GRE score, you should never have the excuse that you didn't properly prepare for the exam.

UNDERGRADUATE RESEARCH EXPERIENCES

Jim is interested in applicants who have four semesters of research experience. Although some professors say three semesters of research experience is the minimum they want applicants to have, Jim is generally looking for students who have started research in their sophomore year (although this isn't a hard-and-fast rule for him). There are a number of reasons Jim wants students who have started research their sophomore year.

First, doing so really helps them prepare for conducting independent research, like a senior thesis, in a lab. Specifically, the earlier you get involved in research, the more likely you will find someone who is willing to mentor you through the thesis. The senior thesis is critically important to demonstrate that you reach beyond what you did in your major, that you are willing to do more work than is necessary, that you are willing to take risks, and that you are dedicated to pursuing research. It is of course ideal if your thesis topic is in the same area you want to pursue in graduate school because you will become familiar with the literature and methods, but the exercise of doing the thesis is more important than the content. If you have a great thesis introduction written by the time you're applying to graduate school, Jim wants it put into the application. This is because it is a great way to showcase how you write, what you're thinking, and that you understand why you're doing what you're doing. This is a chance to show you understand your research (e.g., why are those your hypotheses), that you are a leader, and that you are thinking independently.

The second reason for starting research involvement early on is that the sooner you start in a lab, the more opportunity you will have to try different areas of research. Indeed, some students in Jim's lab have also worked in family studies or language development labs because they had multiple interests. If the student then decides to change labs, he is excited for them to follow their passion because we all want students to chase what they are most passionate about. This opportunity to explore topics and research opportunities is ideal as long as you are still in each lab long enough to get good letters. Jim points out that if a student is in a different lab each semester, the faculty do not know you well and can't write the strongest letter.

The third reason to start research early is that it increases the odds of securing a good summer research experience. If you are applying for paid summer research assistantships with no research experience, you'll probably be competing against applicants who do have research experience and you'll be at a distinct disadvantage. For example, if you are from Minnesota and headed home for the summer, you are more likely to get a summer research opportunity at the University of Minnesota if you can have a professor at your college recommend you based on your lab work. Even when these opportunities are in different labs, the mentor can still write a letter confirming you can think critically and can test hypotheses. In other words, the research topic can change, but the underlying scientific method is the same and you continue to demonstrate increased training in science. Jim got into graduate school to work with the mentor whose lab he worked in during a summer, so these opportunities can be quite fruitful.

A fourth reason to start research early is simply that it increases the likelihood that you can go straight into graduate school from undergrad. Contrary to my assertion earlier, the majority of the students Jim accepts into his lab as PhD students have not done a year or more of a paid research assistantship (although some have; and he agrees this practice is getting more common). Rather, plenty of students can be competitive if they start research early in their undergrad careers. If you don't get started on research experiences early, it almost guarantees you'll have to do a paid research assistantship after undergrad in order to be competitive for graduate school.

A fifth motivator for early involvement in research is securing good letters of recommendation. Jim says that by working in a lab for a long period of time, faculty and professors have a chance to get to know you better. Securing strong letter writers is also a reason to pursue upper level classes like a grad-level class and to take writing intensive courses, which usually have smaller enrollment, or to take multiple classes from the same professor. Jim says that all letters of recommendation say good things, so he is looking for letters that clearly demonstrate the professor knows the student, has interacted with them, and knows what they are capable of doing. Working in labs for multiple semesters helps you get to know the faculty, meaning the letters are going to be stronger and more personal.

FUNDING YOUR UNDERGRADUATE RESEARCH EXPERIENCE

Applying for funding to support your undergraduate research experience is a fantastic way to demonstrate that you understand that securing funding is part of doing research. Jim makes all students in his lab who work on a senior thesis apply for internal funding at Notre Dame. He said the grants are pretty generous in terms of getting a few thousand dollars to fund traveling to a national conference, like the Annual Meeting of the Psychonomic Society, and to pay for subject compensation.

Taking steps to write grants early in the senior year, at a point when you don't yet have all the data, forces students to put their ideas together in a logical way and identify what the stimuli are going to look like, the independent and dependent variables, and the data analysis plan. Having a funded grant application to submit with your application shows what you've actually done. If submitting additional documentation with an application is not allowed by the program, you can email the grant application to the professor separately from the formal application. You only want to send it along if it's good, though! If you're willing to share it with the faculty member, you must think it's pretty good, and you will be judged accordingly. If your grant is funded, that is external confirmation that the document is in good shape, so use the grant process to help devise a writing sample you can pass along to potential graduate mentors.

CONTACTING POTENTIAL GRADUATE MENTORS IN ADVANCE

You do not want to spend $100 to apply to a program if your identified mentor is not admitting students. The graduate admissions process does not work the same way as undergrad. You could be the most qualified person in the world but not get in if the advisor is not taking students. Therefore, it is really smart to contact people in advance to find out if they are admitting students or not.

Don't start your email informally with a first name or "Hey There." You should not slip into being super informal in these inquires. If that's how you're going to introduce yourself to someone you've never met, that's a weird dynamic.

Jim loves being contacted in advance of the application process. In particular, if you are going to a national conference that is just before graduate application deadlines, that can be a great time to meet face-to-face with potential advisors. He said this is also a great way for faculty and applicants to get a preview of each other. Meeting at a conference ahead of the application deadline is especially useful if the student's application materials are not quite above threshold, but they can have a great chat in person. This creates a personal connection with the mentor before you compete with all the other highly qualified candidates in the applicant pool.

SPEAKING WITH CURRENT GRADUATE STUDENTS

Jim suggests talking to current graduate students during your campus visit. He thinks that emailing graduate students might not be very useful because they are not going to say in an email that they hate their advisor or program. Graduate programs frequently invite top applicants to interview on campus, and when you are there, you can spend time with graduate students inquiring about what the facilities are like, what living in the area is like, whether they feel like they fit in, what they like about the program, what's challenging about the program, and what working with their mentor is like. Although most campus visits organize a time for interviewees to go out with current students so they can speak freely, don't forget that faculty will later ask the graduate students about their opinions of you. Be professional and take the opportunity to talk to everyone broadly related to your area of interest, not just people in one lab, because if people outside the lab give consistent answers, such as they are all unhappy, that's a sign of an unpleasant program.

SELECTING YOUR GRADUATE PROGRAM

Jim suggests using the professors that you know to get advice on where to apply. Depending on your credentials, your professors can help make sure you do not waste your time. For example, they can steer you away from top schools if you are not going to be Ivy League material and come up with some really great second tier places where they know you will succeed.

Jim also points out that you don't have to go to Harvard to have a great experience; you just need to be successful. Go where you're going to be successful rather than selecting a program based on the name of the university or the city you're going to live in during graduate school.

If you don't attend an undergrad institution with famous faculty in your desired field, you can still take advantage of the people you know. Although connections help, many well-known faculty who are looking for strong PhD students trust the opinion of friends at smaller liberal arts schools. Even if your professors don't conduct federally funded research, they did attend an institution that granted PhD's, so they will be friends with those federally funded researchers who are looking for students.

APPLICATION AND INTERVIEW MISTAKES TO AVOID

In Jim's experience, the interview isn't going to get you the job, but you can lose the job during the interview. Although some schools have interview weekends before they admit candidates and some host recruitment weekends after candidates are admitted, some common mistakes Jim advises you to avoid across the board are listed below.

- Don't come to the interview unprepared. If you can't speak intelligently about your senior thesis, then you haven't been pushed hard enough to hone your ability to discuss research.
- Don't dress or behave informally. Don't be the one person who wears a sweatshirt and jeans to the interview. Definitely don't do drugs when you are back in your hotel room (yes, this happens).
- Don't be generic. If you apply to ten graduate programs, you should write ten personal statements. Of course, the parts about your background and motivation can be the same in each statement, but in each one, show that you have put a lot of thought into why you are applying to that specific school. For example, tell prospective mentors what you like about their research agenda and how you see your interests fitting in. Mentors want to know that you've spent some time carefully thinking about their labs and programs.
- Don't get sloppy. The application process can be long and hectic, but that's no excuse for making silly errors out of either ignorance or sloppy editing skills. Just be careful not to make errors out of either ignorance or sloppy editing skills. Jim relayed one story about a prospective applicant who emailed him to ask about a position in his lab and who concluded his letter by saying it has been his life-long dream to study in France (wrong Notre Dame!). Needless to say, his application went nowhere. You also don't want to submit an application to Ohio State that mistakenly says you can't wait to study at Notre Dame with Dr. Brockmole (this happens more than you'd think). These first impressions matter because the faculty are trying to assess you with a limited amount of time and information. Because the faculty member is forced to develop a hypothesis about how an applicant is going to develop over the years, sloppy materials or informality up front is a bad sign.

CHAPTER 3

How to Interview With and Select a Graduate Mentor

by Geoff

In Chapter 2 we guided you through the process of building the strongest possible application to a PhD program. This chapter assumes you have been offered interviews at multiple PhD programs. The goal here is to help you successfully interview for those PhD programs and select the best graduate mentor for you.

THE MENTORSHIP MODEL

Do you think the title of this chapter should be, *How to interview with, and select, a graduate program* instead of a *graduate mentor?* If so, you are not alone. When you were in high school and applying to colleges, the question was, "Which of these universities and programs will best suit my needs?" The game is now far more precise. Although you will be admitted to a PhD program in psychology or neuroscience , you are typically assessed based on your fit with one specific faculty member or perhaps a couple of specific faculty members with closely related research interests. You should pick a program based on your interest in working with specific people, not programs as a whole. Ultimately, you will be working in the laboratory of your graduate mentor as well as taking a few classes during the first couple of years, with almost all of your last 2-3 years spent in your mentor's lab exclusively doing research.

This advice applies to master's programs as well. If your plan is to earn a master's degree to help get into a PhD program, it would be ideal to pick a master's program based on whether or not there is a person in that program who fits your research interests. Later, this will help you get into a lab that does similar research when you apply to PhD programs, assuming that is your goal.

The reason for this precise matching of graduate student to faculty member is that academia generally uses a mentorship model much like the apprenticeship tradition of artisanal craftspeople. You are paired with a scientist-mentor. That person teaches you what they know and how to do the work by having you do the work alongside them. Being able to conduct science is a skill that typically takes about a decade to learn and do well. So working intensively with a single scientist, or a tight circle of scientists in a lab, is vital for you to learn all there is to know.

BECOMING AN INDEPENDENT SCIENTIST

The goal of graduate school is to teach you to be an independent scientist in about 5 years. Research on how we become experts suggests that it takes about 10 years to become an expert at something.[1] So cutting that time in half during graduate school is a difficult goal for any program. The length of time required to accomplish mastery is one of many reasons why most programs want to pair you with the expert that conducts research closest to what you want to do. This pairing allows you to hit the ground running. Later in this chapter, I discuss some programs that do not pair each student with a mentor at the outset, adding to Ashleigh's remarks from Chapter 2, but first allow me to describe why graduate programs are designed to pair you with a mentor, immediately at the beginning of your training.

Let's consider all you will need to master to run your own lab in five years. You need to read the scientific literature to figure out what topic you want to study. You need to learn to design logical and rigorous experiments to study that topic and learn what common confounds to avoid in those tasks. You need to understand analysis software, statistical tests, and what inferences can be drawn from the tests. You need to understand the peer review process well enough to estimate whether your findings are publishable. You need to learn the ins and outs of scientific writing. You need to know how to address other scientists' questions who are experts on the topic when getting reviews back on our papers or grants. You need teaching and mentoring philosophies to guide you in training your own students. Finally, you need to balance your focus on research with your teaching duties, raising a family, and so forth. This is actually only a subset of what you will need to know how do when you are an independent research scientist.

The point of telling you how much work is involved in becoming a scientist is not to scare you, but to communicate that you have a lot of learning ahead of you. The mentorship model of academia is designed to accelerate this learning process so that you can leave graduate school prepared to run your own lab. As discussed in the next chapter, it has become common for people to do further training in a postdoctoral position after graduate school so that they can acquire more experience with the skills and knowledge in this long list (see Chapter 5). However, a postdoctoral position is not always necessary. Depending on your field, succeeding in graduate school may minimize the need for a postdoctoral position, so let's get you set up for success.

BEFORE THE INTERVIEW

Before you interview potential mentors you should determine who you'd most like to work with. Evidence shows that interviews can add noise to decision making instead

---

1 Crossman, E. R. F. W. (1959). A theory of the acquisition of speed-skill. *Ergonomics, 2*, 153–166. doi:10.1080/00140135908930419

of increasing the signal (i.e., rather than provide useful information that moves you toward the optimal choice, it moves you farther away from the optimal choice). Because of this, it is wise to go into your interviews with a priority list based on fit and purely objective data.

DETERMINING THE BEST FIT

After receiving multiple offers to visit schools for interviews, the first step to figuring out which graduate school you want to attend begins online. You began this process in Chapter 2 when you applied to graduate schools and looked for faculty that were doing research that interested you. Now you go back online to research the potential mentors who have offered you an interview with greater scrutiny. Download and read papers from those faculty members and decide who writes papers that you understand and does the most interesting research. Although you did this to get an idea of who to apply to work with, you should attack this with greater vigor now. Being passionate about what you are working on is critical for you to find success. From a time management perspective, this reading will be easier because you are now only researching the faculty who offered you an interview. So read a lot and follow your heart, gut, or whichever organ you use to decide whether you are excited about something.

To be competitive for graduate school, you are almost certainly currently working in a lab either as an undergraduate or as a paid RA. As stated in Chapter 2, the easiest path into a lab is to target labs that are doing research in the same general area as the lab you are now in (e.g., your current lab studies decision making, so you would look for other labs studying decision making), or use the same research method as your current lab (e.g., your current lab uses functional magnetic resonance imaging, or fMRI, to study decision making, so you would look for a lab that uses fMRI to study a topic that might interest you even more than decision making). Pursuing labs that investigate the same topic or use the same method will increase your *fit*. *Fit* refers to the match between your potential mentors research program and your interests or skills. Fit is the most important aspect that faculty consider when they are looking at applicants to take into their labs (but see Julie Golomb's interview at the end of the chapter for a different take on fit).

When reviewing applications, faculty members ask themselves whether the applicant has already been introduced to the topic or already knows how to use the equipment. The reason that faculty look for applicants that already have overlap is because of the tall order outlined at the beginning of this chapter: the faculty mentor needs to turn you into an independent researcher in a few short years. If you would rather be studying something else (i.e., a different topic and with different methods), you may need to volunteer in a lab doing that kind of work before you are competitive enough to have your pick of labs for graduate school. This is your life—spend it doing something you really enjoy. Alternatively, you may have to accept an offer from a lab that is not quite positioned on the cutting edge. This is stepping on the toes of Chapter 2 somewhat, but if you have applied to graduate schools and did not get a strong response, a lack of research experience on the topic you want to study in graduate school is likely the reason, assuming you followed the advice in Chapter 2 to protect your GPA, get the best possible GRE score, and get as much research experience as possible.

Let's assume that ranking mentors based on research interests narrows your preference to a few programs but doesn't result in a clear top choice. This outcome is likely because if you have followed our advice, you probably have interviews with several faculty mentors who study similar topics (e.g., visual attention) whose research programs only differ slightly. In addition to research interests, there are a few other useful ways to rank programs before the interviews that will help you make the best decision.

PUBLICATION RECORD

One piece of objective data is the potential mentors' publication record. This means finding out how many papers the mentor's lab published each year, on average, during the last 3–5 years. If the lab publishes about one paper every other year, and already has a couple of graduate students and postdocs, then the likelihood is low that you are going to go into that lab and publish enough to end up at a top research institution. The publication record may be low because the mentor doesn't have enough ideas, space, or funding to increase their capacity to mentor graduate students like you.

Another important factor is the impact of the papers the lab does publish. It is possible that they only publish one paper every other year, but each of these papers is published in one of the top journals (like *Nature, Science, Cell, Proceedings of the National Academy of Sciences [PNAS], Neuron,* or journals that publish longer papers like the *Journal of Neuroscience*). A journal's quality is measured, in part, by an **impact factor**, which represents the average number of citations earned by papers published in that journal. You can look up impact factors online if you are unsure about the impact of the journals where your potential mentor publishes. If the lab publishes infrequently, but in the best journals, then you might do very well in that lab (i.e., even if you only publish one high-impact paper every couple of years).

Talk to the faculty member you are currently working with about the publication records of the prospective labs and ask them where they would go if they were faced with your options. Discussing the mentors' publication rates with your current mentor is also vital because publication rates are different across different fields and methods. If you are working with nonhuman primates, then one paper per year is an excellent rate. If you are running a lab that publishes data from behavioral experiments with college undergrads, then one paper per year is a very low rate.

GRANT FUNDING

The other important consideration that Ashleigh briefly discussed in Chapter 2 is the funding situation in the prospective labs. Do they currently have grants? You can look this up on RePORT (the **National Institute of Health's [NIH]** website for research activities, https://report.nih.gov/) or NSF's website (https://www.nsf.gov/). Read the abstracts of the lab's grants and see how much funding the labs have. When prospective graduate students ask me what my lab is going to do over the next five years, I know they haven't done this homework. If they had checked RePORT, they would know about the research that my lab is funded to perform during the next handful of years. They would also know this by simply looking at my CV, which is posted online.

Why does grant funding matter? A well-funded lab is going to have money to send you to conferences, pay for subjects and equipment, and teach you how to get your own grants. If two potential mentors have similar publication rates and you find their research similarly enticing, you should give preference to a well-funded lab over one that has not been funded for several years or never at all. Grant funding is tight these days, so having a gap in funding is more common for top labs than it used to be; however, if a lab has gone several years without grant funding, this could be a sign that the potential mentor might not excel at teaching you how to write your own grants.

PERSONAL REPUTATION

There is one other conversation that you should have at this point with your current faculty mentor, or other faculty in your department. Ask them if they personally know the people you will interview with. Are they nice people? Are they known for being difficult or obnoxious at conferences or as reviewers of papers? The academic world is small and typically your current mentor will know which people are difficult at an interpersonal level. There are not many difficult people running around the halls of academia, so don't become overly concerned that you are going to end up with one. Most academics are generally introverted and a bit socially awkward (i.e., they are my people), so there are very few tyrants among us. However, there is nothing wrong with asking your trusted mentor if your prospective mentor is supportive, particularly of minorities or women in the field. Indeed, this is a question we wish was queried more frequently because we need more minorities and women in science. Do your homework with your current mentor and other faculty in your department. Academia is a small world, and someone probably knows the person you are thinking about working with.

INTERVIEW TIMING AND TRAVEL

Graduate interview season in academic departments begins shortly after applications are due, and applications are typically due around the beginning of December. Although some people might be brought out early, most departments schedule their interview weekends during the same times as one another in late January through early March. This may mean that you may be invited to interview at two schools at the same time. Do not simply turn one down due to a scheduling conflict. Just let them know that you already have an interview scheduled on that weekend and ask if you can come at another time. This is very common. Faculty members and departments are used to having these scheduling conflicts and will accommodate people coming in at different times. You should go to all the interviews you are invited to because having an interview doesn't mean you will have an offer. So, don't rule out any opportunity because the interview weekend overlaps with a school you might like a bit more at this point.

Are you going to have to pay to travel to these interviews? Typically, no. Healthy programs will have funds from the institution for recruiting and to fly you out for an interview if you are in the country (i.e., at a school in the United States and applying to a school in the United States). Some will put you up in a hotel, while others might have you stay with a current graduate student. When you are invited for an interview, most of the programs extending an invitation will tell you that your travel costs are covered. If they do not volunteer this information, then ask them if they do cover these costs. If the costs are not covered, it might be a sign that the program is not particularly well funded, which should factor into your decision if you have more than one offer.

Unfortunately, it is common that institutions will not cover the travel costs of candidates who are overseas. Some might pay for part of the costs, and sometimes students can fly once and interview at multiple schools in the same country, so that the schools can split the costs. As you can tell, you will need to inquire about what is possible if you are not already in the United States. If you are overseas you should take advantage of Skype or other video conferencing software. It is common for excellent overseas candidates to get offers after only Skype conversations with a couple of faculty (i.e., without an in-person interview). If you are not already in the United States, you are not at a significant disadvantage if you can use Skype to talk with your potential mentor.

PREPARING FOR MEETINGS

Do your homework before you go to your on-campus interview! Begin by looking up all the people in your field (e.g., cognition, perception, behavioral neuroscience) that you hope to study, and the people on your itinerary who you will interview with, as described below. This is the single most important way to prepare for an interview. You will have already looked over the webpages and papers of your potential mentors, but now you want to see what other topics are studied in the program and what methods people use. Whichever school you pick for your PhD, you will learn about everyone's research in greater detail from departmental talks, so hopefully you find those topics interesting too. But the real reason to do this research is to demonstrate that you prepared for the interview by seeking out the knowledge that was available. Scientists need to have inquisitive minds, and if you are not inquisitive enough to Google the department and faculty, then people may judge you as lacking the essential internal drive for knowledge.

Before you leave for the interview, get a list of the faculty that you will meet with. Look up one online and at least skim some of their papers. Some of them will be out of your area. For example, you may meet with clinical psychologists even though you hope to work in a rodent lab. Even if their research seems to be very different from your research, do your homework and know what each faculty member you'll meet studies and how they study it. Following the interviews, all the faculty members will provide their input on you when they meet, so be prepared to impress every one of them. If you do not do your homework on the research of the faculty you are interviewing with, it not only suggests that you might not be inquisitive enough to succeed in grad school, but it also implies that you are not very serious about going there. Also, it's basic social psychology that people are more apt to like you if you appear to like them. Doing your homework expresses interest in them and increases the odds of receiving a positive evaluation.

Doing your homework is also the advice you should receive before you go to an interview if you pursue a job in industry or academia after getting your PhD, so you may as well learn this practice now and apply it to graduate school interviews. If you show up at Tesla corporate headquarters for a job interview and ask them to tell you what it is that they do in this shiny building, they will view you as dangerously uninformed and unprepared. This mistake is not one you want to make when interviewing for graduate school or any other position. But, if you were go into Tesla headquarters, sit down, and say that you've looked over their current set of products and see an opportunity to develop a new kind of vehicle to capture a new kind of consumer, then even if your idea is impossible or crazy, they will be impressed that you've come with an idea and thought about it before showing up. Doing your homework about faculty you will interview with can help show that you have the capacity to seek out information without being told where to find it (although perhaps you are now being told this by me) and to teach yourself.

There is one other thing that you need to do to prepare for these interviews. I ask two questions of people I interview. The first is, "Do you have any questions about the research that we do in my lab?" You just read about how to be ready for this one. The second is "Do you know a programming language?" We are all looking for graduate students that know some computer programming when they start. In many labs, Matlab has become the standard because of the many existing functions that help people program experiments and analyze data. If you have not programmed or modified code to run experiments and analyze data, or taken a class on it, then you should start right now. All object-oriented programming environments are pretty similar (e.g., Java, Matlab, Python), so don't worry too much about the specific language in the course. But learning some programming is the single most important thing you can do for your success throughout the rest of your career. Some clinical or social programs will like it if you know a statistics package like SPSS (Statistical Package for Social Sciences) or SAS (Statistical Analysis System). With a few exceptions, almost everyone wants a candidate that can step in and program right away. When was the last time you were a subject in an experiment that used paper, pencil, and a stopwatch? Probably never.

WHAT DO I WEAR TO INTERVIEWS?

Most of the nerdy people like us who make it to graduate student interviews do not have an extensive collection of business attire to pick from. However, you are going to need at least one nice outfit. There are a couple of times in your career when you are going to need to wear a suit or a professional outfit. Your graduate interview is one of these times. Do not show up in a T-shirt and jeans on interview day. The faculty might dress casually, but you can't. If you want specific suggestions, men should wear a suit and tie. Women should wear a pantsuit and blouse. It's better to overdress than to be too casual. Ashleigh and I would both prefer to risk being overdressed, for example by wearing a blazer, because you can dress down once you're on campus by taking off the blazer, but there's nothing you can do to become more formal if you show up in a collared shirt and everyone else has a suit jacket. Your interview attire will not go to waste because you will be using it again when you interview for faculty jobs or jobs in industry after you complete the PhD program. In addition, women should avoid wearing too much makeup or jewelry. Everyone should wear comfortable shoes because you will do a lot of walking.

Typically, the interview weekend will include a social gathering at the end of the main marathon-like interview day. It is okay to dress less formally for this. Obviously, you should not wear clubbing attire, even if the social event is at a club. Interviewees could wear nice jeans and a sweater to the social gathering. Still avoid items like ripped jeans, flip-flops, higher heels than you would wear to the interview, and revealing clothing (e.g., low necklines). Keep in mind you are on the clock even when people are standing around drinking cocktails and chatting. Have fun and get to know the people at these gatherings, but keep it together at all times (see Dr. Brockmole's interview in Chapter 2).

THE SCHEDULE

Since we already spoke about the faculty meetings on your itinerary, let's discuss other scheduling details. Graduate school interviews involve a packed schedule. You should receive a draft of the schedule before you leave for the interview. These itineraries emulate the type of schedules used when people interview for faculty positions and when faculty go to other universities to give talks. They are very full and you do need to go to everything.

If you are coming straight out of an undergraduate program, it probably seems like a bit much to have the department demand a 12-hour day from you. Indeed, when you are in graduate school, you will be able to allocate your time as you see fit, outside of classes, lab meetings, and departmental talks. However, many of us put in 60-hour weeks more than once in graduate school, so perhaps the nature of these interviews is not completely out of the realm of the effort /you will be asked to put forth should you be admitted to the program. Note that the people in the department are trying to decide if they want to spend the next half decade with you, over a quarter of a million dollars on your training (think tuition, salary, and benefits for up to 5 years), and countless hours helping you get your PhD. This is a big investment in you, so do your best and understand that the physically exhausting interviews are because of the nature of the exciting period you are about to begin.

You'll normally have a longer block of time with the faculty member who you are primarily interested in working with so you can get to know this person. This is the time to ask them about their mentoring style (see the section below about mentoring styles). However, do not spend all your time grilling them. It is usually best to begin by selling yourself. Tell them in a concise and clear way about the projects you have worked on. What was your role? Did you collect the data, perform the analyses, or write it up? Did you present at a conference? Were the projects published and were you an author? Tell them about your programming experience if you have some. If your undergraduate or research assistantship mentor worked with you on learning scientific writing, also discuss that. You do not need to prepare slides. You should simply be able to explain what you've done clearly and concisely with words. It would be good to practice this 10-minute "elevator talk" with your roommate or significant other. Your current mentor would also probably like to hear it once you have practiced.

If you know what kind of mentor would fit you best, then it is good to ask questions to determine the potential mentor's style. You may know what kind of supervision you prefer from jobs you had during or after high school. Did you flourish under tight supervision in which your boss was constantly nearby to give advice about how to handle issues (i.e., told you what to do)? Or did you find the most success under bosses that gave you leeway to figure out your own way to solve problems and simply checked in occasionally to see how you were doing? If you know what you want, then ask the necessary questions to determine if the potential mentor supervises in a manner that suits you. You can also ask existing graduate students about the level of oversight, as I will discuss more in a later section. However, if you do not have a preference, then these questions are just going to add noise to your decision-making process. In that case, it is best to make decisions based on the aforementioned objective data of publication rate and impact.

When interviewing with your prospective mentors, it is good to ask them certain questions that are relevant across all types of mentors. How do they determine authorship? What is the support for conference travel? What are your explicit expectations (e.g., a paper per year, running with my own ideas)? There are no right or wrong answers to these questions, but it is useful to know as much as possible about the environment that you will move into.

MENTORING STYLES

One style of mentorship relies heavily on senior lab members (i.e., fellow graduate students or postdocs who have been in the lab longer) to train new lab members. However, it is always possible that the lab is not big enough to have this kind of intergenerational training.

The apprenticeship model of mentoring new graduate students means that there are many different styles. Data to tell you that any one style is the best do not appear to exist. The data that I am aware of suggest that faculty who think they aren't very good mentors are actually judged to be the best by their protégés.[2] This may be because they try harder to compensate for what they perceive to be weaknesses in themselves. I don't believe that data exists to tell us whether these satisfaction surveys predict career success. However, you will find that some mentoring styles suit your personality better than others. This is something that you will have to determine for yourself. For example, I was not a fan of micromanagement, so having mentors who turned me loose with the help of other people in the lab was ideal. You might want more guidance, in which case hands-on styles might be better for you.

Some mentors prefer a training scheme that puts you on your own project right away, as in my example above, with the expectation that the entire research group will be available for questions and guidance, rather than having you initially work under the guidance of a specific senior grad student or postdoc. This can be an effective strategy too.

Some mentors are hands-on and prefer to meet with graduate trainees several times each week, teaching them how to collect and analyze data themselves. This is obviously necessary if the lab is small and does not have other grad students or postdocs to help. I know people who have used this style of running a lab effectively, even with a fairly large number of lab members. It has the upside of insuring that you are getting information from the highest-level expert in the lab. The downside could be that your mentor does not have enough time to work with you and handle other responsibilities.

MISREPRESENTING YOUR QUALIFICATIONS

I need to mention one issue that I hope is not necessary to mention, except that I have seen it arise. Do not misrepresent your qualifications. For example, saying that you know Matlab or Java when you really do not, is a statement that will get you in trouble. Do not say that you have read a paper or have a skill that you do not. Just be open and honest about what you do know and the skills you do have. Be quick to say you don't know something if you do not.

You cannot fake your way through graduate school. If you misrepresent your skills and you receive an offer to start a program, the expectations of you will be far higher than you can reach. Never do this. People want to know the real you, not some inaccurate characterization of what you wish you were. Having the confidence to say what you don't know is actually a trait that many people look for in potential graduate students. We want

---

2   Godshalk, V. M. & Sosik, J. J. (2009). Does mentor-protégé agreement on mentor leadership behavior include the quality of a mentoring relationship? *Group & Organizational Management*, 25(3), 291–317. doi:10.1177/1059601100253005

you to come to us when you get stuck or don't know something. Students who lie their way in will quickly get kicked out.

POST-INTERVIEW

THE OFFER PROCESS

After you have had your interviews, there will be a week or two during which the faculty deliberate. Then they will make an initial set of offers to prospective graduate students. The offer will state the nature of the tuition coverage, how much the stipend is, and how long financial support is provided, given good standing in the program. The deadline for making virtually all grad school offers in the United States is April 15th.

Although a first set of offers are made in each program within a week or two of the interviews, the process plays out over a fairly long period of time. A minority of programs may make you an offer before a recruiting visit, although this is increasingly rare. This does not mean they are more excited about you than anyone else, but that it is just how they do the recruiting process. In contrast, some offers are made as late as early April. The reason for this is that initial offers might be made to one group of applicants; some of those applicants will have multiple offers and accept one, turning down the rest well before the deadline. When someone turns down an offer, it opens up a slot for the department to offer to someone else. The nicest thing that you can do for the health of the field and the faculty who made you offers is to make your decision as soon as possible so that other young people can be brought into the field. No one is served when people wait until the last minute to finalize their decision.

A quick comment on who sets the number of offers a department or faculty member can make: the administration that pays the bills will tell the department how many new students they can accept, and usually how many total offers can be made. This is based on the amount of funds available in the college that houses the department. Then, the department internally decides how to allocate those resources among the faculty. The goal of these decisions is to establish equity across faculty. For example, if one faculty member has five students, and another member has none, then the member with none will likely be allocated one or two offers to make. The availability of grant funds clouds this simple math a bit, as having students on grants often allows an individual faculty member to grow a lab larger than a faculty member who only has institutional funds available (meaning the college pays for the students). If your offer is not funded by a grant, you will typically serve as a **teaching assistant (TA)** each semester in exchange for your salary from the college. The reason I am telling you this is so that you know why a potential faculty member might say that they are not recruiting students or may be unable to make you an offer even after you interview. Do not take it personally because often faculty interview people that they would like to take, but the scarcity of slots prevent that faculty member from being able to admit everyone that they would like to admit. Because of the many moving parts involved, your potential mentor may not be sure if they have a slot when they interview you, but hopes they will be able to make an offer if you impress them.

The slowly evolving offer process means that if you do not have an offer by mid-February, you may still get one. If you have not gotten a rejection letter from a department by the beginning of March, this is great news and means that you may yet get an offer. What does not help is to email the faculty and department staff in the interest of lobbying them to make you an offer. These decisions are made at a group level and lobbying here doesn't work like it does with congress. If you want to work on a backup plan, begin looking for paid RA positions, a common pathway into graduate school (as suggested in Chapter 2). During your time as a RA, you can learn how to do research, improve your computer programming, and publish. All of this will make your application much stronger the next year.

MAKING YOUR FINAL DECISION

Some graduate school candidates will have offers from multiple schools to choose from. Congratulations if you are among them! If you do not have such choices, do not despair. You only need one offer if it is the right one. Out of the mere seven schools to which I applied, I only had one offer, plus six rejections, and one school that never even sent me a rejection letter. This is the lowest possible level of success at that stage that can still result in obtaining a PhD. However, I have subsequently been invited to give talks at many of the schools I did not get into and have even had faculty job offers from a couple of those schools. Getting rejected by an institution at one point in your career does not mean you can't someday be a professor there down the road. But, if you do have a choice between multiple alternatives, then let's consider what matters and what doesn't.

THE MENTOR

The most important thing to consider is the mentor you would work with at each school. Is one more productive? Do you find the methods or questions being asked by one more interesting than the other(s)? We already talked about how you should use the publications of potential mentors to establish an initial priority list. After the interviews, you will know more about what they are doing and the direction things are going in their labs. Use this information to create a new priority list. Have there been any changes in your priorities? If so, why? Was it the beautiful weather the school happened to have on that weekend, or the lighting in the building, or was it really about the research you would be doing and the resources available to you?

THE PROGRAM

The specific details of each individual program largely don't matter. Some programs will have slightly more classes required than others, some might have a fellowship year, grad student housing, or a slightly larger stipend, given the cost of living. These details can make a slight difference. In particular, the amount of your stipend can be important for quality of life in graduate school, but the truth is that most of this stuff will have a very minor impact on the magnitude of your success. Being a TA or instructor is valuable experience for a later faculty position or running your own team in industry, so trying to avoid being a TA is like trying to avoid eating your fruits and vegetables. Obviously, a higher stipend might mean you can live more comfortably (although it probably also means a higher cost of living), but that benefit only lasts for the years you are in graduate school and the real goal is to get your ideal job, not afford more pizza for a few years.

Classes don't matter in graduate school nearly as much as research productivity, as I discuss at length in Chapter 4. So if you need a tiebreaker, then you should slightly favor the program with fewer course requirements. I advise my graduate students to spend as little time on classes as possible and as much time they can in the laboratory. I note that this perspective about the importance of classes may be controversial. Ideally, we would like students to be great at everything. However, if you are faced with capacity limits, then decisions like this may be necessary.

LOCATION

The location of the school is not important, assuming your work-life balance and mental health are not negatively impacted by the geography. In fact, it would be ideal to work with a mentor who is at a school in a city you do not want to end up in. You are going to be a product of that school. This means that they are going to want you to go out into the world and conquer it. You are not going to be invited to stay as a **tenure-track** faculty member after completing your PhD. In the rare circumstances in which people have returned to take a faculty position at their PhD institution, it was typically following a period in which the person left, did a postdoc or had a faculty position elsewhere, and was hired after years of great success elsewhere. This means you don't pick a graduate program that is in your dream city. Target that dream city for a faculty job once you are in graduate school working with your dream mentor. Plus, being in an attractive location might actually hurt your productivity. Think living near the beach is appealing? Me too, but it would also keep me from working in the lab when the sun was out and the surf was up.

Graduate school is only about five years. By its very nature it involves deferring gratification to a later point in life. Embrace this. I feel it is necessary to point this out because I know people who have picked graduate schools because they wanted to be close to their family or be in a sunnier place. Then, the 4–5 years will fly by and after a postdoc, you must solve a harder problem. That harder problem is getting a faculty job at a place that you might want to be for the rest of your life. My advice is to look at graduate school as a tool that will enable you to get a job anywhere you would like. Going to graduate school in the place you would really like to spend the rest of your life is not recommended. It makes it harder to succeed in graduate school and when you finish you will be forced to find work elsewhere. Graduate school is not the final destination.

NAME-BRAND SCHOOLS

Does the name brand of the school matter? That is, if I publish 10 papers in graduate school, would it matter if I have a degree from Fancy-Ivy-League University or the University of Middle-America-State? This is a somewhat controversial question. Ashleigh and I had success working hard and publishing well at a University of Middle-America-State. However, it is true that if all other things are exactly equal in terms of your productivity and other qualifications after graduate school, you will likely have a slight edge in future hiring decisions if your degree is from a prestigious private university. The reason for this is that fancy private schools spend a lot of money to poach the best people from other institutions so that their brand remains good in the minds of academics. The quality of the faculty is one of the factors that keeps their rankings high.

But do not think that getting a degree from an elite private school is sufficient to ensure career success. You will still need to be as productive as students from anywhere else. Hopefully your school will have the facilities and resources to help you succeed, and by that, I mean plenty of grant support and equipment, so you can spend more time collecting data than peers at other institutions.

Let's say your choice comes down to two mentors both with grant funding who study topics that you love. The mentor at the fancy name-brand school has a full lab, a tight schedule, and can only offer you limited time on the data collection equipment because lots of other people in the lab want to use it too. Alternatively, the other mentor is at a slightly lower ranked school with fewer graduate students, so they can offer you more attention and better access to equipment. You will likely be better off picking the latter. Finally, keep in mind that even the best schools have faculty who are less productive than some of their peers at less prestigious institutions. Conversely, some schools that are not universally known as powerhouses have faculty who are world renowned in what they do. This is a good topic to talk to your current mentor about; what would they do if given your choices?

CONTACT GRADUATE STUDENTS OF POTENTIAL MENTORS

Because the mentor with whom you are going to work is the most important factor in your graduate school success, some applicants contact previous graduate students to see how they liked working with the faculty member. This is smart. However, you will likely be contacting the successful ones (because unsuccessful students will fall off the radar), so take these enthusiastic responses with a grain of salt. Similarly, asking the existing students, postdocs, and RAs how they like working in the lab can be useful. But again, keep in mind that this information is not always completely reliable, because some people provide feedback about experiences that are more about them than about the lab in which they work. People who are struggling will see plenty of blame to go around. Successful people will likely say the environment is perfect. Still, it is useful to ask those individuals to explain what it is about the lab that they think has allowed them to succeed or makes working there particularly difficult.

Here is a list of questions you could draw from. If you are talking to the person face-to-face, then you could probably cover all of these without it seeming like an interrogation. If you are going to send someone an email, or talk to them on the phone, you might want to be more selective and just pick a few from this list. Like detectives on police dramas, you really want to just get the person talking and they will paint a picture for you.

- What is morale like in the lab?
- How are lab meetings run? Are they run like journal clubs or are they used for students to present findings and trouble shoot new experimental designs?
- Do you have regular meetings with Dr. Evil?
- When you joined the lab, was there a senior person in the lab who helped you get oriented? If not, how did you acclimate to the lab?
- How would you describe Dr. Evil's mentoring style—is she more hands-on or hands-off?
- How promptly does Dr. Evil respond to your questions and emails?
- How long does it take to get a manuscript back from them?
- Do you feel like you have the support that you need from both Dr. Evil and the department?
- What do you wish you knew before you started this graduate program?

DECLINING AN OFFER

Keep in mind that the field is small and even declining an offer needs to be done with the same standard of professionalism that you exuded during the interview process. I usually find that when I get a series of emails from someone to whom I've made an offer about the nature of my mentorship, or the weather in my city, it is because they are trying to justify a decision. The decision-making literature suggests this is a common mode of information seeking, known simply as the *justification effect*. If you already know what you want to do, just go with that decision. However, if you have a legitimate question that you are curious about, then go ahead and ask it. I simply point out that taxing your potential mentors with questions that are not diagnostic is unwise. You will regularly run into the faculty members who made the offers you declined at conferences throughout your career. These encounters will be awkward for you if you had asked them a bunch of bizarre questions (e.g., what kind of tree they would be if they were a tree or where they see themselves in 10 years) before sending them an email to notify them that you are rejecting their offer for graduate admission and will be attending a different school.

ENJOY THE RIDE

Interviewing for graduate school can be a very exciting time in your life. Much like interviewing for faculty positions across the country, it is a time when you are trying to predict what will happen if you follow this path or that one. Allow me to pass along the advice of Gordon Logan, my colleague who looks like Gandalf with slightly shorter hair, but who has less cryptic advice than Gandalf. Gordon points out that each student's success will almost completely be determined by how hard they work. Different mentors and programs will point you in different initial directions, but whether you move down that path quickly, slowly, or at all, will be determined by how hard you work and how much you think about the papers you read.

Ashleigh's interview with Julie Golomb, PhD

faculty.psy.ohio-state.edu/golomb/

Julie Golomb, PhD is an Associate Professor in the Department of Psychology at The Ohio State University. She has affiliations with the Center for Cognitive and Brain Sciences, Center for Cognitive and Behavioral Brain Imaging, Neuroscience Graduate Program, and Neuroscience Research Institute. We wanted to interview Julie for this chapter because she has had success working under top mentors in the field to earn her PhD and as a postdoctoral research fellow, as well as success recruiting strong graduate students and postdocs to join her lab at Ohio State. As you will see, Julie is a thoughtful, caring mentor who advocates for others in the field, especially those she has mentored.

ON THE ISSUE OF FIT

Julie agrees that fit is crucial in choosing a mentor, but thinks of this factor a bit more broadly. For example, Julie thinks it's less important that a student comes into the lab having previous experience working on the same exact topic or technique as her own research lab. Rather, Julie looks for students who have a specific interest in her research topics, even if their previous research experience is with a different area of cognitive psychology or neuroscience techniques. The reason Julie does not require that an applicant have experience in the exact same methods or topic as her is because she believes that as long as students have a solid research background in the fundamentals (experimental design, data analysis, computer programming, critical thinking, et cetera), they can learn the more specific skills in her lab. So rather than putting all the emphasis on a particular methodological background, for example, Julie has a different take on fit. Julie's thoughts are rooted in the concern that a student may feel pressured to apply to labs that are closely related to the type of research experience they've had previously, simply because they believe they have the best chances of getting into a good program that way, not because they are genuinely interested in that area. How does Julie determine someone is genuinely interested? She says she looks for signs like a student having read her papers, asking critical questions about the research, generating interesting research ideas, and exuding excitement when talking about the topic.

WHY INTERVIEWS MATTER

Julie's perspective on the purpose, method, and utility of the interview process encompasses the faculty's perspective and the applicant's perspective. For both the faculty member and the applicant, the other has already crossed some threshold of suitability on paper. A main purpose of the face-to-face interview for both people should be to get a sense of interpersonal chemistry and research compatibility. From the faculty perspective, Julie explains that accepting the student is a serious commitment. This perspective shows you what a great mentor Julie is because she really believes that her commitment to her students extends their entire career. Julie wants to be their advocate and support them for a lifetime, so she emphasizes the need to mesh well interpersonally. Even for mentors who don't take this long view, most of them will agree with Julie that given how much time a faculty mentor spends one-on-one with a student in discussion over the next five years, if their personalities and research styles are incompatible or if either of them feels like they can't communicate effectively with the other, it's a red flag.

Beyond the interpersonal chemistry described above, Julie thinks that interviews serve a much bigger purpose for the applicant than for the faculty. While the faculty member is solely evaluating the applicant, the applicant is evaluating the faculty member, the lab, the program, and the overall research (and living) environment. For the faculty member, the interview could potentially be just as effective over Skype, but for the applicant there are many more boxes to check during an interview, many of which she believes are best done on campus if possible. Julie agrees that the objective metrics Geoff mentioned above (e.g., publication rate) are the most important in deciding on a mentor, but she also adds some other factors that the student should consider during the interview. First, she agrees that you shouldn't make the decision based on beach access, but she says you do need to make sure it's a place you would feel comfortable living for the next five years. Second, since you will sometimes be spending 60 hours a week or more in the lab, your lab mates will become your family. This means that you need to make sure that you can willingly spend a huge portion of your days, almost every day, with the people in the lab and in the lab environment.

SELECTING A JUNIOR VERSUS A SENIOR MENTOR

Julie raised the very important issue of choosing between a junior and a senior mentor. Generally speaking, a junior mentor is a faculty member who does not yet have **tenure** and a senior faculty mentor is someone with tenure. Julie knows that some people advise that the senior mentor is always a safer bet. However, she disagrees. A senior mentor is not always the safest bet, and in fact in many cases, students benefit more by working with junior people. She described a risk analysis whereby the benefits of a senior mentor tend to include experience mentoring (and placing) lots of students, more prominent status in the field, more stable funding, and sometimes more flexibility for students to pursue their own research ideas. However, these benefits aren't always present. It's also possible that a senior mentor may have a lab that is so large that each trainee has little time with the mentor. The senior mentor may have increased administrative responsibilities and may take a more hands-off, impersonal approach to advising.

Julie notes that there can be benefits of working with junior faculty with whom things tend to be more personalized. The advantages of a junior mentor tend to include more time mentoring their students hands-on, providing more close one-on-one training, having fewer graduate students in the lab, and perhaps less competition for lab resources (and the mentor's attention). Other benefits might be that a junior mentor is potentially more ingrained in the research and able to provide stronger technical training because they more recently trained with the latest techniques themselves, especially in fields like fMRI, where methodology and equipment develop rapidly. However, Julie points out that being someone's first graduate student is a bigger risk because you're probably going to be the guinea pig as they develop their mentoring style, set up the lab, and figure out how to train and manage students. So, although you might get more attention, there might also be more growing pains. For example, in trying to get the lab up and running the mentor might confront struggles with equipment and construction, or might not have grant funding for conference travel and subject compensation. Further, although she probably has **startup** funding for a few years, if she doesn't get a grant quickly, money might run out before you've graduated, so they will probably be more protective of the startup funds, which could impact your research. All that being said, junior faculty do tend to be hungrier for publications. This is important because they will be looking to publish as much as possible and publications are just what you need to establish your name in the field.

Finally, Julie thinks that if you are looking at a senior faculty member, it's fair to factor where their recent graduates have gone into your decision (i.e., have they gotten jobs and where are they employed?). However, she says that for a junior faculty member, the sample size is just too small for that information to be useful.

Julie creatively tried to address potential risks of working with either a junior or senior faculty mentor by suggesting that you pick an institution where you have one primary person you want to work with but the department has other people with whom you'd also enjoy working, collaborating, and interacting with. This can be especially important because it decreases your risk in case a lab is lonely (e.g., if you're the only graduate student or if the advisor ends up being a poor-quality mentor, regardless of where they are in their career).

QUESTIONS TO ASK YOUR POTENTIAL MENTOR

Julie suggests asking your potential mentor the following during the interview process. She cautioned that there might not necessarily be a best answer to some of these questions, but the answers would arm you with more information to help you decide between programs.

- How do students choose a project? Are they handed an existing project when they start in the lab?
- How long does it usually take people in your lab to graduate?
- What's your mentoring style?
- How much access will I have to everything I will need to collect and analyze data? For example, if it's an fMRI lab, how much time will I have with the scanner?
- Will you have money to send me to conferences? What's your policy on going to conferences? For example, do you have to have the paper written before you can submit an abstract? Will you pay for me to attend a conference if I'm not presenting?
- Is my funding dependent on the department or the lab? If the lab, do you foresee having funds for the next five years?

THINGS YOUR POTENTIAL MENTOR WISHES YOU KNEW

I love asking faculty what they wish the interview candidates knew. The mismatch between what applicants know and what the faculty think they should know can be very telling. Filling in that gap is part of why we wrote this book! Julie said that it is important that applicants understand what a typical academic lifestyle is like and that it may not be for everyone. Julie's experience is that a lot of undergraduates expect that graduate school is like undergrad. They assume that graduate school is focused mostly on coursework, or think that because they are good at managing their classes and getting A's, they will be successful and happy in grad school. Julie says that students often underestimate both the commitment and intensity of being a grad student, and that this is a critical reason why students should make sure to get an immersive research lab experience before applying to grad school. Volunteering in a research lab for 8 hours per week will not give you the same sense of research life as a full-time RA position. She recommends treating full-time summer research programs and RA positions as a practice run for grad school, to get a feel for if this is what you really want. Julie says that a lot of students don't understand just how much graduate school takes over their life, but the all-encompassing nature of the training can be an asset if you love research and the lab.

Finally, Julie points out that most students applying to graduate school are accustomed to being successful. In undergrad, being smart and hardworking will get you the A. But in research, there is a lot of failure you have to be able to deal with because, for example, you won't get every grant, every experiment won't work out, and manuscripts get rejected from journals. The flip side is that graduate school will be more than just classes, in a good way, for students who belong in graduate school. Julie says that your lab really should become your family. It really is the case that you're joining a group of people with whom you may spend time over the weekends and may build lifelong friendships, and your advisor can be like a lifelong friend, rooting for you throughout your entire career. Julie finds that grad school is often more challenging than students think, but also a more fun and rich experience than they expect.

CHAPTER 4

How to Succeed in Grad School
by Geoff

So you got into graduate school. Now what? You might think that getting into graduate school is the hard part. However, many people who get in do not transition well to the demands of graduate school once they arrive. The problem is that the skills you acquired as an undergraduate to ace tests and memorize textbooks are not the skills you need to be successful in graduate school.

Not only are there virtually no exams in graduate school, but as we've said, classes and grades are not the major criteria on which your success is judged. Instead, success in graduate school is defined by how strong a CV you can build by publishing papers, gaining teaching experience, learning to write efficiently, having experiment ideas, and other goals that are 180° from the measures of success that apply when you are an undergraduate. For some first year graduate students this comes as a shock. Tenacious independent problem solving was probably not something that you learned in class as an undergraduate but it is going to be the most important thing in graduate school.

WHAT IS GRADUATE SCHOOL?

The goal of a graduate program in any field of science is to produce independent research scientists. This means that the faculty are trying to get you ready to not depend on them. Typically, your training will change across time such that you initially receive a great deal of help and oversight from your faculty mentor, both of which will wane across time.

As you pass through the different stages of your graduate training, your mentor and graduate program will expect that you take ownership of your research program. Although you begin by doing what your mentor asks you to do, your mentor expects you to increasingly identify the next question to be answered or the next experiment that needs to be run in your line of work.

No first year graduate student feels like they will be ready to run their own laboratory when they complete their graduate training in four or five years. Indeed, few of us were. In most fields of science, postdoctoral research positions, sometimes several of them, have become the norm (as I discuss in Chapter 5). A postdoc is a position that you have after getting your PhD, in which your salary is paid from a research or training grant. These positions allow you to spend 100% of your time on your research, acquiring the remaining skills you need to run your own lab. I say this so that readers will not be paralyzed by fear at the idea of being the head of a lab or a faculty member in just a few years. The typical path to being an independent research scientist passes through multiple top-tier labs in which you acquire additional skills that insure your success when running your own lab in academia or industry.

In what follows, I describe how to be objectively successful in graduate school. What this means is that you leave graduate school with the papers and skills needed to be highly successful in academia or in industry. If, on the other hand, you are interested in a career that can keep you employed without the stress of running a research group, the advice here should still help your application for any position rise to the top of the pile.

HOW SUCCESS IS MEASURED

Your success in graduate school is measured by how many papers you publish and how impactful those papers turn out to be. Those papers are built on the data that you collect during long days and nights in the laboratory, including weekends and sometimes holidays. You will learn how to analyze data, write a scientific paper, and respond to reviewer comments to publish those papers.

Let's say you publish 20 papers in graduate school, including a paper in a **glossy magazine** (e.g., Nature, Science, or Cell), but at the same time your grades in your classes are all a B or B-. The great news is that you will still have your choice of the best jobs after graduate school. Authorship on 20 papers is about as many as I have ever heard of anyone publishing in graduate school. This is a spectacular accomplishment that is nearly impossible to reach, and will make it certain that you will be highly sought after in the job market. In contrast, chances are those grades will never be seen again.

In graduate school, a letter grade of a C is equivalent to an F and the class would need to be retaken in most programs. This means that the grading scale primarily involves A's and B's, and almost everyone gets an A or B in each of their graduate classes. This restricted range in the grading scale renders grades almost completely irrelevant, which makes sense because grad school is all about research and publications anyway. It is possible that you will want to get a job that emphasizes teaching when you get your PhD. In this situation, building a strong teaching portfolio is critical, but you still need to complete research so that you have a dissertation to defend and several publications to explain how you earned your PhD, which is after all, a research degree.

One of my friends from graduate school likes to say that the job of a graduate student is to be a data machine. I find this to be a concise way of describing the primary task in graduate school. That is, collecting data is most of the job. It is important to note that the data machine also becomes a machine that spits out scientific writing and experiment ideas. So you will need to read and think about papers. Go to talks and think about how you would test the ideas described. However, these latter products of your graduate training come along naturally after a high volume of data has poured out of the machine. A paper that came out several years ago showed that people who got faculty jobs at the top research universities published about 2–3 papers per year in cognitive psychology.[1] This means that by the time you are a postdoc or junior faculty member you should be trying to publish about three papers per year. In certain fields of neuroscience, such as the study of nonhuman primates, the goal is closer to one paper per year. Your mentor can give you a good estimate of publication rates among recent competitive job applicants in your field. Those are the rates that you should shoot for if you'd also like to be at the top of your field. Next, I will describe how to make that possible.

GETTING YOUR FEET UNDERNEATH YOU

What to do on day one? The first thing you should do when you arrive in your graduate program is to meet with your graduate mentor and ask what experiment you should start preparing to run on that day. Often this means helping with data collection or analysis on a project that another graduate student, or your mentor, has already started.

Helping with an ongoing project was my first taste of research in grad school. I was tasked with programming and running follow-up experiments for two projects that a third year graduate student in the laboratory had begun at least a year prior. This gave me someone to ask specific questions of, who also had a vested interest in getting me going. This senior graduate student was more available than even the most hands-on mentor could be. In addition, it was much less nerve racking to go to a fellow graduate student or postdoc with a question when I was stuck, than it was to ask my mentor a question that I feared might not be a great question.

The other lab members are there every day, almost all day. This means that during your first months in graduate school, you will learn more about how to become a data machine from other lab members than you will from your faculty mentor. This is not because your mentor doesn't like you or is shirking responsibility. Rather, faculty members at research universities have a ton of responsibilities in addition to mentoring grad students. They have to teach classes, edit journals and review papers, serve on many committees of various types that keep the university operating, and many also serve in other positions like chair of a program, Director of Graduate Studies, or even dean. All these responsibilities pull your mentor away from the laboratory, even though they would prefer to be there with you, watching the data come in, and enjoying the excitement of scientific discovery.

---

[1] Byrnes, J. P. (2007). Publishing trends of psychology faculty during their pretenure years. *Psychological Science, 18,* 4, 283–286.

WHEN SHOULD YOU START ACTING ON YOUR OWN IDEAS?

Everyone I have ever known comes into graduate school and wants to prove that they belong there. This is because most of us feel that we are not as smart as the people around us. This feeling is natural. The other people in the laboratory or department have spent years learning the jargon that we have difficultly decoding and turning into English in our heads.

The idea that we are not as smart as the other people in graduate school, and science in general, is known as *imposter syndrome*.[2] I believe there are probably psychopathic individuals who never have these feelings in graduate school. However, the vast majority of students have these feelings when they start a program. Sometimes the feelings of inadequacy can last for years, even in highly successful graduate students who go on to great things in their careers. So if you have these feelings, you are normal.

When you learn the jargon used by the other people in your laboratory you will feel more comfortable. As I mentioned above, this can largely be accomplished by reading papers that your lab mates refer to, as well as virtually any other paper in your area of research that you can get your hands on. Think about each paper and how it relates to the other papers you have read. Look up terms you don't know, and the papers cited in the papers you read. In time, thoughtful reading will make you feel like an expert; impostor syndrome will fade with each paper you read and integrate into your memory.

Some people want to prove that they belong in graduate school by going rogue. They want to have an idea that turns into a new discovery and a big paper. This is exactly what you should do as a second, third, or fourth year graduate student. If you are precocious you might have a successful idea in your second year. But *do not go rogue for a couple of years* after starting graduate school. Some of the ideas you have early on might be good. Write these down in a notebook and keep them for later. However, it is more likely that you are thinking of something that has already been done.

Why is coming up with your own idea and developing your own experimental design as a first year graduate student unwise? First, you do not know the literature well enough to know what is publishable. Your general idea might have been done already. For example, there might be an entire area of research that is described using different terms than you are using, explaining your mentor's, peer's, or even the Internet's inability to assess the novelty of your idea. Second, the people around you have spent decades learning science. Start by utilizing their knowledge. Learn how to use the equipment and design the kind of experiments that are run in the lab before you try coming up with your own. Do something in the wheelhouse of the laboratory so that you can learn how science is conducted. After learning everything you can about how science is conducted in the laboratory, then you can innovate. Essentially every revolutionary figure in the history of science or art, such as Darwin, Curie, and Picasso, first became an expert in their domain,

---

[2] http://www.apa.org/gradpsych/2013/11/fraud.aspx

learning what people had done before them. Without extensive background knowledge, you are likely to spend a lot of time reinventing what has already been created and using terms that are inappropriate to communicate with your colleagues. Communicating your ideas is one of the hardest things scientists do, and you need to learn the vocabulary before you can convince the field to change how it thinks about things.

As mentioned previously, get a project from your mentor, perhaps straight out of their grant. That research needs to be done and is a good enough idea that the grant reviewers funded it (a high bar these days). Start running that as soon as possible and thinking about what you would do as the next step. Talk to your mentor about thoughts you have for that next step. Develop this line of research as a starting point.

The metaphor that I like to use is to start out with a laser beam focused on a specific research question. Then, as you learn the literature necessary to understand where that beam is focused, you can broaden the beam to cover a larger and larger area until you have multiple projects in the same theoretical neighborhood.

DON'T COLLECT DATA WITHOUT HAVING YOUR ANALYSES WORKED OUT

Always collect your pilot data first, and analyze that data set immediately. Like many people, I would run myself through the experiment to see what I was asking subjects to do. Then I analyzed those data. With certain time-demanding methods of neuroscience, this usually requires running someone through the entire experiment. Why do you need pilot data to analyze?

The most common pit that we fall into as graduate students is collecting data without knowing if they are worth a damn. In my field, almost all experiments are run using computer code that we write ourselves. Then we analyze the data using code we write. This leaves many opportunities for mistakes. Once you get your program running, it is seductive to begin running that experiment as fast as you can, assuming that you can analyze the data after you've collected an experiment's worth. However, I have often made a programming error that makes the data unanalyzable and worthless. Collecting data with that program can result in wasted months or even years.

This is so common that I can provide many examples from my limited experience. One scary example of this was when someone I worked with forgot to include the command at the end of the file that wrote the data to an output file. After collecting what should have been three experiments' worth of data, they went to analyze it only to find out that they hadn't collected any data files at all. I worked with someone else who programmed and ran an experiment with 45 subjects, with each subject completing two 4-hour sessions. Data collection alone was a 360-hour investment. They were in a hurry and had not performed their analyses beforehand, checking to make sure they had the right number of each type of trial. This meant that they didn't notice a fairly large programming error that made the location of the task-relevant information predictable, even though the point of the experiment was for it to be unpredictable. You need to know if you are really running the experiment you intend to before you set off running it as fast as you can.

In addition to catching simple errors, frequently, after you have collected pilot data, you realize that your experiment is not designed appropriately to do what you want. Plot the data. Look at the graphs. What are the critical comparisons that you need to test the hypotheses, and do they exist in your design?

If you look at your data and they do not make sense to you or the other people in the lab, then after ruling out a programming error, there are several other possibilities you can assess. The data may be noisy, indicating that you need more samples to measure the phenomenon of interest. Or the response you are trying to modulate might not be detectable in your experimental set up. In the latter case you need to back up and get the known effect first, then add your new twist to it.

I am not advocating that you run statistical analyses after collecting each sample to look for a result early in the process and become strongly biased to see a certain result as you collect more data. Recent methodological commentaries have warned that confirmation bias can result in findings that other people cannot replicate, and thus, do not stand the test of time. If you think you can't provide unbiased instructions to subjects, then you may need to have someone else instruct the subjects and maybe collect the data. This other person should not know how the results are turning out or what the hypotheses are.

Another option that does not violate guidelines for the responsible conduct of research, while not screwing yourself by neglecting to work out your analyses before you collect the real data, is to run your experiment twice. The first time, you collect a set of data that you analyze, make sure your scripts are working, and that you have set up everything correctly. Then, you can use this initial dataset to feed a power calculation to determine how much data you will need when you run the real experiment. This is a rigorous way of performing a power calculation for your experiment even when you are doing something new and don't yet know the effect size to expect.

The danger of collecting data with a flawed design and only finding out after several months that you were not really ready to collect data is real. Most of the people that I have seen make this kind of protracted mistake have difficulty recovering from them and completing the projects they started. After doing it once, they get panicky and are even more rushed to run something that works the next time.

My advice is to collect pilot data from a sample of a few subjects. In different areas of neuroscience this might mean collecting data from a couple of cells or tissue samples, or collecting a couple of days' worth of data from the first animal or first few patients. Then, analyze your data so that you can see the results from the pilot sample. Does it look like you are measuring what you want to? If not, think about whether your experimental design has enough power. You might need a stronger manipulation so that your effect can be more easily measured.

Another major reason you might not be seeing what you expect is that you are making a mistake in your procedure. Go over what you are doing with your mentor to make sure your methods are sound and have a reality check about whether what you are observing makes sense. Knowing what observations or pattern of results are impossible requires someone with a lot of knowledge in the field, and this is what your mentor is for. Do not be afraid to bring them preliminary results that do not make sense to you. It is their job to help you figure it out. I find these to be some of the most interesting problems to solve in my lab, and these situations allow me to use the expertise I spent decades acquiring to help my grad students.

It is always possible that you are simply not seeing what you expect because you are doing something new and because variability is inherent in all measurement, which is why science uses averaging to get a better estimate of the true signal embedded in the noise. In this case, you might just need to collect more data and see if some kind of signal emerges from the noise.

Scientists are just people and all people sometimes make mistakes. It is far better to discover your mistake after a week of work than after a semester of work. The former is not a big deal. The latter can be demoralizing and have you questioning whether science is a good field for you. Mistakes are common. We have all make them. The difference between the successful and the unsuccessful people is how quickly they can identify mistakes, recover from them, and correct them.

WHY HAVING MULTIPLE PROJECTS GOING AT THE SAME TIME IS ESSENTIAL

Above I introduced the idea that prejudging an experiment, in which you root for a given outcome, is a dangerous thing. One way to avoid this is to set up experiments that test mutually exclusive predictions of competing hypotheses. For example, the ideal experiments are those in which one hypothesis predicts one pattern of results and a second hypothesis predicts a different pattern of results. However, another powerful way to inoculate yourself against dangerous bias is to maintain multiple projects.

Once you have learned how to collect and analyze data in the lab, you need to develop multiple lines of research. Why?

There are two primary reasons that having more than one line of research is necessary. First, not all experiments or even entire lines of research work out. That is, sometimes we design an experiment and our results cannot be interpreted, even though we have sufficient power and a logical experimental design. Years later you might discover why.

Almost all researchers I know have run an experiment or set of experiments that yielded a pattern of results that could not be explained in any coherent way, and thus were not published in a peer-reviewed paper. This becomes a scientific dead end. When it happens on the only line of research you have going, it can be demoralizing. Worse still, if a dead end is reached after a couple of years of working on a single project, students may be tempted to consider making ethical compromises, what my kids call bad choices, in which scientific fraud might seem like a possible course of action to salvage your career. If a dead end occurs a couple of times in a row, a number of honest people decide that they are not lucky and that they are poorly suited for science. When you keep a couple of lines of research going at the same time, these common setbacks are not catastrophic. Instead, you simply shift your focus over to your other project

In graduate school, approximately a third of my projects were dead ends. Some of these were not due to uninterpretable data, but because the findings did not seem novel or exciting enough to publish. Ashleigh had better fortune, with nearly all her projects ultimately published. Since you never know what the result of an experiment will be, multiple lines of research give you projects to fall back on when things don't work out.

The second major reason why multiple lines of research are necessary is that it prepares you to have your own lab. The goal of graduate school is to train you to be an independent research scientist running your own lab in academia or industry. Independent labs are always conducting multiple lines of research.

Think of your mentor's lab. It has a couple of graduate students, each with their own projects, and perhaps a postdoc and research assistant with their own lines of work. Graduate school is a nice, safe environment to hone your skills of thinking about how different puzzle pieces fit together and working on different parts of the puzzle simultaneously. Of course, no one ever really does two things at the same time. Instead we shift between working on one for a while, then shift back to the other when we need to pause for any variety of reasons. For example, when you submit a paper on one line of work you need to wait for reviews to come back. This period is perfectly suited to switch over to your other line of work instead of simply sitting idle for a month or two with white knuckles.

As we discuss in later chapters, not only should you have multiple projects, but it is also ideal to have a diverse portfolio. Most of the time, you will want to maintain a line of research that is almost guaranteed to work, but unlikely to completely change how we think the brain works. This kind of project is usually a follow up on previous work in the lab or the field. This is a good, safe kind of project to have going. You should have some more high-risk stuff. It is good to have a project that seems like a long shot but would be big news if it worked. This could be developing a new task or devising a new method for your research. It could also be testing a wild prediction of a model you have thought up. These high-risk, high-reward projects keep life interesting. When they pay off they are big news, and when they do not, it is not a big loss because you have your other less flashy lines of work to fall back on.

WHY YOU SHOULD VOLUNTEER TO GIVE LOTS OF TALKS IN GRAD SCHOOL

Raise your hand if you love public speaking! Not you? That's how I felt too.

When you get your PhD you will be expected to be able to lead groups of people in doing research. This is true whether you are planning to stay in academia or go into industry, government service, a nonprofit organization, and so forth. Being a doctor is cool. But with that coolness comes the responsibility of explaining to other humans what you are doing and convincing them that what you've discovered is important.

You are going to spend years in graduate school learning the jargon that allows you to talk to other experts in the small, specialized group of researchers that study the same topic. This allows you to talk to them efficiently. However, when you give a talk to the broad listenership of your department, you need to describe your research using simpler terms that can be universally understood by nonspecialists in the audience.

Giving talks also allows you to practice several other skills that will be essential as your career progresses. You need to be able to describe the background that motivates the work. Why is your method ideally suited to answer the question? What are the competing hypotheses and why do they make different predictions? These points are easy to gloss over when you are talking to your advisor or other people in your laboratory. Making these aspects of the logic explicit is exactly what you will need to do in papers and when you give job talks in the future. So while you are in graduate school, one of the best ways to develop these skills is to give talks. You can give talks at lab meetings, to your department, or even in university-wide forums. Obviously, you have to already be a data machine, cranking out cool results to have something to talk about, but once those cool, new data start flowing, you are ready to start practicing how to tell the world about them.

Giving a talk in your department can seem scary. But this is a safe environment for tuning up your speaking skills. The people in your department and area of research are trying to help develop your skills. In addition, the stakes are low. If you try out a new way of describing your research and people find it hard to understand, then you've learned a valuable lesson without losing a job offer.

When you get your PhD, the expectation is that you will be able to teach people. When you give a talk on your research you are really teaching people about what you are doing. In academia, you will be expected to teach classes about other people's research. By giving talks on your research you are developing the skills you need to teach. When you give a departmental talk, you should imagine that undergraduates are in the room and you want to make them understand what you are saying. If you only practice giving talks with the top people in your field as your target audience, then very few people who are not experts will understand the material. Always try to reach everyone, and then sprinkle in some nuggets for the specialists that they will find particularly interesting, even if those denser nuggets might be a reach for the undergrads. Ideally you provide enough background and information about the methods that even undergrads feel like they understand everything, maybe for the first time.

You might think that these speaking skills will only be useful if you become a professor at a university. This is not true. If you go into industry, then a likely expectation is for you to be a leader, particularly if you want to have a chance of promotion. If they wanted someone who could just use the equipment they could get someone with a Bachelor's or Master's degree. If you become the head of a research group in a company, you will need to lead meetings, give presentations to boards and executives about what your group finds, and even talk to the public about your developments and discoveries.

When students start giving talks they often fear the questions that will be asked. This also happens when we start teaching. Questions are not bad. Actually, questions are the best part. They reveal when you've lost people, assuming the question is something like, "Could you walk me through the experimental design again?" When the questions are more advanced, such as what the relationship of your work is to other theories or what the results of other manipulations might yield, it means that the audience is following you and you are stimulating them to think about connections and next steps. This means you are crushing your talk. No one will generally tell you that your talk changed how they were thinking about the topic, so you typically need to determine this by the level of the questions you receive.

RULES OF THUMB ON GIVING TALKS

Allow me to give you a couple of quick rules of thumb on giving talks; you will get more of these from your mentors.

Rule #1: Less is more. Do not overload your talk will a ton of stuff thinking that people will be impressed by the huge amount of information you are dumping on them. Only give them the most relevant background needed to understand the experiments you are going to tell them about. Ask yourself, "What do they need to know for the next slide?" Ideally, you should only have to show the audience one or two background methodological slides to understand what you are doing. No one wants a 45-minute survey of everything that has been done on a topic or with a method before finding out what the point of your research was. Instead, get to your data quickly. A famous person in our field once joked: that if someone doesn't present their own data on their second slide, he'll walk out of the talk. His point is that he doesn't want a lecture on what other people have done; he's coming to the talk to hear what the presenter has done. When Ashleigh and I are listening to talks at conferences, we will sometimes jokingly whisper "If there isn't data on slide #2, I'm outta here." We don't really mean that, and usually it's because we need more coffee, but you get the idea.

Rule #2: Use common English words when describing your question, methods, results, and implications. Tell them what the question is using English, not a bunch of jargon. Tell them about your method and the measures. Again, do this so simply that someone from the history department would get it. Actually, a good test is to ask if you would understand the talk after taking just a general psychology class. If not, then go back and replace the scientific jargon with common words that a general psychology student could understand.

Rule #3: Don't present the findings of 10 experiments in a 50-minute talk. A talk is not the time to try and get credit for every experiment you've conducted. Present one experiment clearly. If you have a couple of others that build on the same methods and questions, then you could put them after the first, clearly explained experiment is presented. Research shows that people think you are smart if you make it sound simple.[3]

Some people have personality characteristics that genuinely interfere with their ability to give a great talk and to feel comfortable speaking in front of people. I assume these people would not feel better even after practicing extensively. However, I can say that I thought I was such a person and found that the fear of speaking in front of people faded over hours behind the lectern.

BECOMING STRICT WITH YOUR TIME MANAGEMENT

In our modern age of information technology it is easy to become distracted. Do you sit down at your desk in the morning and first check several news sites, scan social media, and read a couple of your favorite blogs? If the answer is yes, then you have probably spent an hour or two doing a bunch of nothing. That is wasted time that does not advance your research agenda.

One of the problems with graduate school is that you are given huge amounts of unstructured time to complete your research. It is very easy to get into the habit of using that time to do things that do not move you closer to your goals. A great strategy to avoid problem that may arise if you have too much unstructured time is to schedule your time strictly and stick to that schedule. This is what you will need to do as a faculty member or research scientist, so it is good to develop vital time management skills now.

As undergraduates, not all of you were so busy that you needed a calendar or appointment program on your computer to keep your schedule straight. But this becomes absolutely essential in graduate school. It is a good idea to formally schedule your writing

---

3   Oppenheimer, D. M. (2005). Consequences of erudite vernacular utilized irrespective of necessity: problems with using long words needlessly. *Applied Cognitive Psychology, 20*, 2, 139–156.

time, data collection time, the talks you want to attend in the department, and so forth. This strategy makes you immune to the problem of walking into the lab at the crack of noonish, sitting down at your computer to surf the web for a while before trying to figure out what to do that day, and then talking your lab mates into going to get a coffee.

Because data collection is the most important task you do, it is best to start with scheduling that. Schedule your time on the equipment a week or two in advance and reserve as much of time as you possibly can. Most labs have rules to make sure all lab members have access to equipment. Use all your time on the equipment that you can schedule. Scheduling data collection weeks in advance will make sure that you don't find yourself getting too busy to collect data. String a couple of wasted days together and the data machine will have shutdown.

Most books on scientific writing recommend a similar strategy for developing as a writer. Specifically, several of the best books recommend scheduling 45 minutes to 1 hour of writing at the beginning of each day. I find that the writing juices have a good flow in the morning, as many professional writers have noticed throughout history. It may sound crazy to begin scheduling time for writing in your first year. However, you are already working on your first experiment, which will need to be published as a paper. Writing the introduction section of the paper will help you understand what you are doing, even as you are just designing the experiment. You can make a figure illustrating the experimental design and write the method section while you are programming and preparing to run the experiment. Then you have the introduction, methods, and Figure 1 for the paper done before you even finish collecting data. When I have the time, I still prefer to begin working on a paper before designing and running the experiments to help clarify the ideas instead of designing an experiment based on some vague jumble of ideas free floating in my mind.

Here is a little trick for efficient time management. Ask yourself if you are doing something that will end up being on your CV. If you go to talks, besides learning about a new topic of research, you typically go with the goal of having an experiment idea that will end up on our CV. Teaching class, grading papers, serving on a committee? All go on the CV. This may sound like a brutally utilitarian way to view the beautiful process of becoming a scientist. But once your CV is in good shape, you can loosen up a bit. Spend more time going to talks that are a bit of a tangent to your main interests. Maybe you develop a hobby or two to keep your mind from feeding on itself and going crazy. However, when I was first starting out it was necessary to develop fairly strict time management skills so I didn't get lost in a sea of unstructured time. Here I provide an example of what several days looked like from my graduate training. As you move through graduate school, your time allocation will shift, and you need to be adaptive. For example, when you are writing your dissertation, your days will be focused on that task, and I found I had to stay out of the lab to do it or I would get distracted by the comparatively fun task of data collection.

You need to reevaluate your time management occasionally. When I was a postdoc, I found myself looking at news sites at the beginning of each day. After thinking about how much time I was investing in this, I trained myself to not open my browser at work except to look up papers. I also found that I was spending too much time reviewing papers. This is something that I still have to keep an eye on because I am prone to say yes to paper reviews and find myself spending too much time on a task that is important but does not help strengthen my CV as much as writing papers does.

HOW TO DEAL WITH HARD TIMES OR QUIT GRADUATE SCHOOL

In my experience, approximately 50% of people in graduate school decide that being a scientist who runs a lab is not a great fit for them. For many people, this is not an easy decision. Some students seek a PhD to prove something to themselves or their parents. They may have dreamed of being a doctor for years. Others might have undertaken their studies with the idea that they really liked being an undergraduate and wanted to do some more schooling like that. Arriving at graduate school and finding out that it is very little like being an undergraduate can be very disappointing.

It is possible that you will find that your mentor's advising style does not suit you well. This could be because you feel like you are struggling, but they insist that you are fine, and that you do not need to change anything that you are doing. Or perhaps their management style is too tight for you and you feel like they are breathing down your neck every day. The first thing you need to do is ask for what you need. If you need space to work, then politely tell your mentor that you think you would be even more productive if you could have a day or two to work on a task before meeting together and providing a status update. Or if you feel lost and you are not sure what you should be doing, you might need to ask your mentor to meet with you to talk about how your experiments' findings are coming together and the next steps. Addressing these kinds of issues might help develop a better working relationship between you and your mentor.

It is also possible that your mentor is set in their ways and cannot change how they give you advice and oversight. There are two options here. One would be to find another lab with which you could collaborate on a project. You could pitch this to your mentor as a way to work together on something that blends everyone's expertise and makes everyone more productive. This would offer an opportunity to see how things work in another lab. This strategy would give you some idea of whether you have expectations for your mentor that are unlikely to be satisfied by someone else. This is akin to having a sleepover at a friend's house so that you can find out whether your parents are as crazy as you suspect. Over time, you could transition to spending more time in the new lab and less time in the lab that does not suit you well.

The other option is to look at the other labs in your department and find another one to work in. Your graduate studies director will be useful in this regard, and it is also up to you to do your homework to figure out who is doing research that you think will interest you, and to which your skills will be a benefit. Although these changes of scenery do occasionally result in a student finding success, you are far more likely to find that you do not fall in love with research in the new setting either. If this is the case, then there is nothing wrong with a more radical change of direction, as I discuss later in this chapter's section on dealing with hard times.

Part of the problem is that, as Jody Culham notes, grad school is not really school and not really a job.[4] It is more like a scientific apprenticeship program. You are being mentored to become a practicing scientist using a tradition that lacks standardization across the field. It is also one of the hardest jobs you will have. It is mentally demanding and requires a huge time commitment. Some days in the lab might be more than 12 hours long.

During difficult times, your goal may switch to not going crazy while becoming a scientist. Throughout this book we will discuss how to kick butt in a lab, a field, and in your life. That means striking a balance between a healthy obsession with a career and taking time for ourselves so that we can take healthy breaks to maintain our sanity. But sometimes in life we decide that a situation is not good for us. At those times the question is how to move on with our lives. Before we get to that, I advocate for a waiting period.

Many people have brief periods during graduate school and during the other phases of a scientific career, during which they wonder whether they have made a mistake and should really be doing something else. I have friends who are highly respected and accomplished leaders in their field who spent a few weeks in graduate school contemplating whether they would be happier doing something else. I believe this is natural. Most of the people I know who have gone through these existential crises find that if they take a week or two off to consider their choices, they come back rededicated to grad school and the career. If you take a couple of weeks and don't miss it, but instead are completely convinced that it's killing you, that could mean that science is not for you.

Some of those who decide science is not for them want to complete the PhD program with a plan to work in industry or even some other goal that does not involve using their research skills. However, a significant proportion of people decide that they really want to leave graduate school and go in another direction. If you leave graduate school, you can typically obtain a Master's degree as a parting gift, assuming you leave after about two years in the program. The precise number of credit hours and requirements needed to get a terminal master's degree vary from institution to institution. You can find out the specifics in your graduate student handbook. The requirements are typically not difficult, and this degree can be thought of as a reward for your hard work in research and coursework during the couple of years you've spend in graduate school.

---

4   http://www.culhamlab.com/academic-advice/

I would like you to keep a couple of things in mind. Your mentor, or your department more generally, does not want to keep you there if your heart is not in it. You might feel like you are letting your mentor down by telling them you would like to move on, but they know this is your career and the decision is about your life. Some people might try to talk you out of it if you are particularly skilled, given what an awesome life we scientists have. Professors and scientists routinely rank at the top of job satisfaction surveys. But they really just want to do what's best for you and if you have already made up your mind about what you want to do, then you should feel confident in your decision.

Some students think that their parents or someone else in their life will be disappointed if they do not become a doctor of philosophy. The truth is that no one else can decide what makes you happy. You have to follow your own truth and do what you love. If it is not science, then you can successfully rule out that option, like a good scientist.

Science is passion driven. It requires you to be excited about doing it because the financial compensation is often not worth the amount of time you spend thinking about your work. Many of my friends and colleagues think about work while they do the dishes, mow the yard, work out, eat, shower, and so on. In contrast, most 9 to 5 jobs do not require thinking about work at home or the gym. If a healthy mental obsession with a science career is not there, it can be difficult to succeed the way that many people want to.

Finally, if you find that you do not have the skills to succeed, such as the ability to talk about your research or lead your own research group, then it is unlikely that you will find great success in industry or in a teaching job. Although PhDs who cannot speak well in front of an audience sometimes work in large organizations, industry expectations are generally that PhDs are able to problem solve, communicate, and lead people. Teaching jobs certainly require instructors who are excellent in front of a group of students. A master's degree may be a good fit for you if you don't think you have the appropriate skill set; the degree would increase your starting salary in a job after grad school.

Now back to telling you how to get that PhD and have your pick of jobs afterward.

THINGS TO DO NOW TO DEVELOP YOUR CAREER FOR THE NEXT STAGE

Although you are still in graduate school, there are a number of things that you can start doing to improve your profile and make a name for yourself. The first couple of these have to do with your online presence.

Your institution will give you an email address. However, it is very useful for you to get your own email address for business use that can change with you as you relocate. Gmail is a popular because it is a well-established service provider that will probably be around for a while. This allows you to have a contact email on your papers that can shift with you as you move between institutions, all the way to the top!

Now for your webpage. You might think that you do not need much of an online presence because people will read your papers and you will get known in the field that way. But the truth is that your webpage is something that you need to hone and maintain throughout your career. It is a place for reviewers of your papers to go to see where you came from. It allows people to find out where you got your undergraduate degree and with whom you worked. As you move toward a job after graduate school, people will go to your website to learn about your work. That job after graduate school approaches quickly. These days, people often begin to position themselves for postdocs, faculty jobs, and industry by their third year in graduate school. So spending a weekend during your first or second year designing your website is time well spent.

One option is to have your institution host your webpage. You can design it using their template or design your own with fairly inexpensive software that gives you complete control (e.g., Everweb, Sandvox). Another option, which Ashleigh prefers, is to have a third-party service provider host a website that you design with their software (e.g., Squarespace, Wix). The advantage of the latter approach is that you can then keep the same website and URL for your entire career. As you move from one institution to the next, your website will stay in the same place, allowing you to build your brand in your field.

Next, begin going to talks with the goal of generating an idea. That is, think about the speaker's question and ask yourself how you could answer that question or test that hypothesis with your methods. Once you spend a couple of years doing this and become an expert on the methods you use, you will often find that you could design an experiment that could address virtually any question. This kind of mental exercise allows you to become the kind of idea factory that you need to be when you have your own laboratory.

The next suggestion is the nerdiest yet. Talk to your fellow graduate students about your research. If possible, talk to your significant other about it. If they are not an academic, better yet, because you will get to explain what you are doing to a nonexpert, which is the ideal audience to help you simplify your communication. This probably sounds obvious or crazy. That is, you may naturally do this, which is great. Alternatively, you might think that you don't know enough about what you are doing to tell someone else about it. But particularly if you are in that second group who is unsure about how to describe what they are doing to someone, then you absolutely need to do this.

Frequently, talking to someone else about your work makes you realize what you don't really know. Sometimes a naïve listener asks a simple question that becomes an essential direction for your next series of experiments. Your mentor already uses this technique. They go to conferences and invite speakers to meet with precisely because these meetings help them talk to people about their research. It fosters new ideas and uncovers gaps in their own logic. Please get into the habit of talking about you research to help you develop your own understanding. If you cannot explain an analysis or a project to someone else, it means that you don't actually understand it as well as you think. We are all prone to the bias of thinking that our implicit understanding of material is excellent when actually, there are some things we just don't know.

Talking to other graduate students about your research has another benefit: it keeps everyone away from using these conversations to complain. It is easy to get caught up in a negativity vortex, which is a whirlpool of complaining that you can get sucked into. Hearing day after day about how everyone is underpaid, that someone's mentor is a tyrant, that teaching assistantship duties are difficult or unfair, and so on, is draining even for the most upbeat person among us. By discussing what you have recently learned or some exciting aspect of your current projects, you can keep your discussion solution-focused. You are also helping the other graduate student with whom you are having the conversation. Research-focused conversations are what you and your friend will need to have with their colleagues in academia, industry, or whatever career path you each end up choosing. So keep in mind that you are helping yourself and also helping your friend practice this vital skill.

Many of us end up making very good friends in graduate school. A good friend is someone with whom you can share actual details of your private life, including the struggles and negative emotions that you have. I am not saying that you should bypass conversations on topics that build these friendships. Instead, what I am advocating for is the skill of monitoring how you are using your time. This is something of an extension of the ideas about time management I discussed earlier in this chapter. Ask yourself if you are developing relationships focused on complaining, or if you are balancing discussions about problems with discussions of your successes, hobbies, outside interests, and interesting theoretical issues you are tackling in your research. I believe that one reason scientists have a tendency to end up in intimate relationships with other academics is that they favor relationships that balance being able to discuss scientific issues with conversations about how to be happy, like what the best restaurant is for sushi, and so forth. We hope that you can keep your discussions with your peers on a similarly useful and balanced plane.

I'll make a final comment about relationships. Neither Ashleigh nor I am a relationship guru who can tell you how to find the perfect partner. However, we know one thing. Do not view your mentor, or faculty in your program, as potential relationship material. Engaging in an intimate relationship with someone in a position of authority over you presents conflicts of interest, potential violations of federal law, and it could kill your career before it even gets off the ground.

You and the people you engage in intimate relationships with need to be at the same level in the organization. Seeing another graduate student is somewhat common. Likewise, faculty members often end up in relationships together. However, sexual relationships between people of different statuses in the field are very tricky (e.g., a faculty member and a graduate student, or a graduate student and an undergraduate). Universities have offices dedicated to dealing with these issues. Society has recently seen an increase in people coming forward with allegations of sexual harassment and misconduct by older people and people in positions of authority. If you find yourself in a situation in which you tell a supervisor that you do not appreciate their behavior and that they should stop but they don't, the university has people employed to help you deal with it. Although you can talk to your department chair or director of graduate studies, you can also go directly to the University Office of Affirmative Action (note that this can go by other names, but you can do a web search for this office by putting in the name of your university and *sexual harassment*). We hope that the knowledge that people are reporting these incidents with a greater frequency will reduce their occurrence, and we hope that it never happens to you.

Returning to the good things that you can do to help your career, I want to emphasize that the most important thing for you to do is publish as many papers as you can in the best journals that you can. There is virtually nothing you can do that will help you more. Publications are what search committees will look at when you apply for faculty jobs. Publications are also what you will use to show your skills when you apply for industry jobs. If you think you want a job at a smaller liberal arts school, then you will want to get strong teaching experience, where you are the instructor of record responsible for teaching your own class. At many schools this is possible after you are an advanced student (e.g., **all but dissertation, ABD**). Find a research topic that you love, publish as much as you can, and get some teaching experience. Grad school will go well if you can focus on those three simple things.

Ashleigh's Interview with Yuhong Jiang, PhD

jianglab.psych.umn.edu/Welcome.html

Yuhong Jiang, PhD is a Professor of Psychology at the University of Minnesota. Yuhong got her PhD from Yale University, and was a postdoc at Massachusetts Institute of Technology (MIT) before beginning her first faculty position at Harvard University. She currently runs a federally funded laboratory that has trained graduate students who have gone on to faculty positions at Brown, Cornell, Delaware, and Dartmouth, as well as top schools abroad. Yuhong has performed significant service to the field, including being an associate editor for an APA journal.

FORMATIVE GRADUATE SCHOOL EXPERIENCE

Yuhong had a very interesting graduate school experience from which she developed a perspective that unified her suggestions on how to succeed in graduate school. Yuhong's advisor relocated to a different institution at the end of her second year in the PhD program at Yale. Despite being invited to go along with the advisor, Yuhong made the decision to stay at Yale. Due to a number of faculty leaving that same year, there were simply not enough labs to absorb all the graduate students orphaned by those faculty, and she was left without an advisor. This experience led her to realize that she was solely responsible for her own career.

Whereas many graduate students have a default attitude that there is a sense of safety because they can rely on their advisor when they need it, she lost that sense of having a strong support person and advocate in the department quite early in her PhD program. Yuhong says that when you know you are singularly responsible for your blunders, you are forced to drop the habit of seeking approval and do everything yourself. Yuhong looks back on that experience with appreciation for learning to be independent, comparing it to the concept that one matures faster in a poorer family. Although Yuhong suggests trying to instill this sense of independence in all graduate students, she recognizes that if an advisor is around to assist students, it's not very effective to instruct students to imagine that the advisor is not around and that they need to be responsible for their own career.

MOST IMPORTANT QUALITY

Early in Yuhong's first year of graduate school, her mentor was asked "What's the most important quality to become successful in academia?" His answer was *persistence*. This answer left a strong impression on Yuhong because it would not have been any of the top traits she would have thought were important at the time. Yuhong says that the reminder of the importance of persistence has stuck with her because she has often needed to be persistent at times when she hasn't had full control over certain aspects of her career.

COMMON STUMBLING BLOCKS

The main problem Yuhong sees graduate students struggle with is the writing process. In other words, they really love research and asking questions. But when they get the answers to their questions, they move on to the next question, without writing up the results from the original questions for publication.

Yuhong acknowledges the lure of crunching data and coming up with new ideas. She said that she believes these particularly interesting parts of the research process make people lose interest once the data are analyzed and the questions are answered. She believes the allure of chasing data prevents people from sitting down to write up their findings before chasing the next set of data.

To Yuhong, more is lost than just not building their CVs when students don't efficiently write papers. Yuhong asserts that writing is a great way to deepen one's thinking and integrate one's knowledge. Yuhong says that even though it's not a preferred task, writing really helps students think about how their data fit with the literature and think about it more deeply. If you see writing as a tool, the first draft doesn't have to be great. Developing your own understanding must happen in order to formulate sound future directions.

A secondary stumbling block is programming. Yuhong points out that it is better to learn how to program early on in graduate school so you don't depend on other people to write the code you need. There are a lot of resources out there that can help you learn. In fact, Yuhong has posted some videos[1] on YouTube that she created for the students in her lab to learn how to code that have over 7,000 hits! She suggests that when you start graduate school, it's a great idea to make the first thing you do figuring out how to code the first experiment you want to run. Yuhong even has the undergraduates working in her lab learn how to code, recognizing that this skill gives them more flexibility and freedom in their work.

DEALING WITH STUDENT-ADVISOR CONFLICT

Yuhong's perspective of independence explains her approach to conflict. She thinks there is a relatively simple answer to dealing with most conflict between a graduate student and their advisor. She believes these situations are due to mismatched expectations between the student and the advisor. Rather than have an attitude of *I depend on this person*, Yuhong suggests approaching the student-advisor relationship as *I'm taking advantage of the resources available to me, including this person, who is my advisor*. If you are thinking of the advisor as a colleague and advocate, but you are responsible for your own graduate career, then most conflicts can be resolved by talking to other people on the project.

Yuhong also points out that to apply for faculty positions, you will need recommendation letters from three to four people. So you want to get to know multiple people quite well. This means that it doesn't make sense to exclusively

---

[1] http://jianglab.psych.umn.edu/VideoLibrary.html

rely on just your mentor all the time for advice, guidance, and collaboration. For example, it's a good idea to have side projects with other people and go to their lab meetings. This is also good practice for a faculty job because in a department, some colleagues will have research related to yours, and these early training experiences will be like those future collaborations.

DISTINGUISHING BETWEEN UNDERGRAD AND GRAD SCHOOL

Yuhong agrees with Geoff that nobody cares about your grades in graduate school. When you look for jobs after graduate school, no one is going to ask about your grades. Yuhong points out that most people who get into graduate school did well in school, but then in graduate school the course load is lower (although the load varies significantly from one program to another) and you wonder what to do with your time. You are still a student, but not the type of student you have been since kindergarten. Rather, the emphasis now is on asking scientific questions and creating new experiment ideas. She notes that an important part of successfully navigating this transition is time management.

TIME MANAGEMENT

If you have multiple projects, Yuhong advises that you don't do them in parallel. When you spread your time too thinly across concurrent projects all of them will slow down. She suggests prioritizing one over the others at any given time and pushing that one to the maximum speed possible. For example, if you need 16 more subjects in the prioritized experiment, use all subjects on that project to get it done rather than divide upcoming subjects among your 5 projects.

Yuhong also suggests that you give yourself a very firm deadline when it comes to writing. It should be short—even unrealistically short. Eventually, your timeline for writing a paper with 3–4 experiments should be on a scale of weeks rather than months, even as little as one week, provided you've read the relevant literature. That time span would mean you'd be putting off many other tasks. Yuhong acknowledges that it's not possible to do much of anything else if you are writing a paper in one week. Yuhong believes that if you don't give yourself an intense deadline, then months or even years could pass without completing the paper. The danger in this is that everything slows down and nothing progresses. The paper may not be beautiful—it may even be a bad paper—but Yuhong thinks it is really important to honor the deadline and get the writing done.

Yuhong suggests that some things might need to be stopped during your writing week because you are thinking seriously and deeply. For example, a critical step when writing a paper is clarifying in your own head what the results of your experiment mean. Polishing the paper can always be done at the very end because, even though it takes time, it doesn't take as much mental effort. I loved Yuhong's approach to writing as a division between two different categories of tasks. One type of task must be done with complete concentration (e.g., writing the rough draft). The other type of task can be done with occasional distractions, or when you have an hour free, like when a subject cancels (e.g., inserting references, polishing figures, formatting). As a parent, this seems like a wonderful approach to many of the tasks parents have to do—some things must be done when you have hours away from the kids to really focus, and others can be done when they are at swim practice and you have 50 minutes to work sitting poolside. Figuring out which tasks fall into which category is a great habit to develop in graduate school.

GRANT WRITING

Yuhong points out that graduate students have many opportunities for research fellowships (e.g., APA dissertation funding). Such programs are geared toward grad students and students should always be on the lookout for grant announcements and submit an application when they see a grant they're eligible for. When writing your first grants you will want feedback. So always write at a pace that allows you to complete your grant well before the deadline to leave enough time for other people to give you comments. For example, you can't send your advisor a draft the day before the deadline and expect feedback in time for you to implement their comments by the deadline.

Of course, applying for a grant increases your chances of financial support, but Yuhong sees another major reason for writing a grant. It also gives you the opportunity to describe your research ideas, which is especially useful if you haven't summarized them in a talk or paper. The grant review committee is often an audience that spans disciplines, requiring you to write the proposal in a way that's understandable to people who are not in your field.

Yuhong says that in her experience there are two types of students. The most common type are students who don't apply for grants and just use the default funding because in most programs funding is guaranteed for 5 years. The other type of student is the one who will apply for every grant they are eligible for. Yuhong says the latter has the advantage because every submission is productive. Although the success rate may not be high, even unsuccessful applications are useful because you get the experience of applying and will have honed your ability to describe your research.

PREPARING FOR LIFE AFTER GRADUATE SCHOOL

Yuhong says that many people start graduate school with the goal of getting a position at a large research university, but many change their mind. These changes happen for reasons such as the low number of available positions or not enjoying the day-to-day life of a scientist. She made a number of great suggestions of what to do during graduate school to stay broadly marketable.

First, Yuhong suggests accepting opportunities to give talks and treating every presentation very seriously. You will learn a lot about presenting your research that will be useful in the future itself (e.g., for conference presentations, job talks, teaching demonstrations). Yuhong advises writing down what you are going to say for the first few slides and the transitions, and practice in front of other people. However, the work you put into preparing is not as much to avoid blundering your way through the talk, but because you benefit from the whole process of preparing. For example, Yuhong points out that practicing talking about your research may help you teach classes someday. This is because most psych classes involve selecting relevant information to teach, especially upper level courses that don't utilize a textbook. This process of choosing what to emphasize and teach involves a lot of the same critical thinking as preparing to give a talk about your research.

Second, Yuhong encourages her students to take additional coursework that might benefit them if they do not stay in academia after grad school. For example, she suggests they take coursework in statistics, data science, or computational modeling. Yuhong points out that we don't have full control over our own destiny, so that even if you are a perfect researcher you still do not have total control over your job prospects. You can't predict or control which universities are going to be hiring in a given year or in what areas they will be hiring. She often tells her graduate students that the graduate program will help train them to read critically, write well, and be generally well-trained in statistics and math. There are many things that this package of skills will do for them. However, she wants her students to pick up skills during graduate school that enable them to be marketable outside academia and she sees taking additional coursework as helping fulfill that goal.

CHAPTER 5

How to Use a Postdoc as a Launching Pad

by Geoff

So you only have a year or two of graduate school left. What's next?

WHAT IS A POSTDOC?

A postdoctoral researcher (known as a *postdoc* for short) is an academic position that you have after getting your PhD, in which your salary is paid from a grant. It's basically a middle ground between graduate school and a faculty position. These positions exist because scientists get grant funding and can spend some of that money to hire someone with a PhD to work on the research proposed in the grant. Sometimes researchers fund postdocs with training grants that are designed specifically to train postdocs to run their own labs. Finally, postdocs can also obtain their own grants to fund their training, as we discuss later.

As scientists, we spend money to hire a postdoc (which is more expensive than funding a graduate student) because we want someone with higher caliber training and skills. The average postdoc salary can rival starting salaries and benefits of junior faculty members at smaller teaching schools. Generally, postdoc positions are year-to-year contracts and do not come with any service or teaching requirements.

POSTDOCS INCREASE COMPETITIVENESS FOR FACULTY JOBS

Most of the top jobs in academia, such as those at powerhouse research institutions, will go to people who have worked as a postdoctoral trainee. There are exceptions to this rule; some people get faculty positions straight out of grad school, but this is a small number of people in psychology and a smaller number yet in neuroscience (e.g., 3–5% of new PhDs according to the Society for Neuroscience).[1] For the other 97% of us who are involved in the scientific study of the brain and mind, a postdoc is an essential part of training. The reason why most of you will need a postdoc to be competitive for top jobs is that you need time to build a CV filled with a strong publication record. If you

---

1 https://www.sfn.org/Careers-and-Training/Faculty-and-Curriculum-Tools/Training-Program-Surveys

skipped Chapter 4, we highly recommend that you look it over to remind yourself about the metrics used to judge academic success.

No first year graduate students feel confident they will be ready to run their own laboratory in four or five years when they complete their graduate training. Indeed, few are. Postdoc positions help bridge the gap between being a student in graduate school and being a completely independent scientist and faculty member. In most fields of science postdoctoral research positions have become the norm. These positions allow you to spend 100% of your time on your research, acquiring the remaining skills you need to run your own lab. We say this so that readers will not be paralyzed by fear at the idea of being the head of a lab or a faculty member in just a few years. Although some people do progress into faculty positions that quickly, the most common path to becoming an independent research scientist passes through multiple top tier labs, as a grad student and then as a postdoc, in which you acquire skills that ensure you will be able to successfully run your own lab in academia or industry.

The first question that you might have is whether you really have to do a postdoc. Couldn't you just apply for faculty jobs during your last year in grad school? The answer to the question is that you absolutely can and should apply for those faculty jobs if you and your mentor think you have a chance at securing one. However, few people are highly competitive for their dream jobs straight out of grad school. If you have worked really hard in grad school to be at the top of the heap, you are probably competitive for a middle to lower tier faculty job.

There is a big caveat here. It is possible that your dream school is looking to hire someone just like you the year you plan to finish your PhD. Because of this large unknown factor, a good strategy is to apply for a set of jobs that you are really interested in at the same time that you are contacting people about being a postdoc in their lab. Using this two-pronged strategy ensures that you have a job lined up when you defend your PhD, and keeps your options open for the potential miracle.

The odds are good that you would benefit greatly from a postdoc. The reason for that is that grad school flies by. Few of you will be truly ready to set up your own lab, begin advising graduate students, write papers and grants as the senior author, and teach full time. The competing demands of your dream job at a fancy **R1** university are difficult to manage, even for those of us who have been doing it a while, so it can be overwhelming for someone coming straight out of graduate school.

I'm assuming you are convinced that a postdoc would help further your training. Below, I cover what you will learn during that time. However, I first want to discuss how to go about finding a postdoc, because this is an ill-defined problem.

FINDING A POSTDOC

There is no standard procedure for lining up a great postdoctoral position. It would be great if there were some central website or clearinghouse for people seeking a postdoc, or for people trying to hire one. There are professional society websites where people post job ads for postdocs as well as a growing number of wikis.[2] However, these ads might not be the best way to find a postdoc that is a good fit for you. Moreover, as a faculty member, I have not found that ads like these help me find the best candidates. Instead, many postdocs find jobs through a less formal route.

The informal marketplace for postdocs occurs through personal communication: potential candidates reach out to faculty and inquire about doing a postdoc with them. Identify major players in your field. Look for people who produce excellent trainees and end up in top jobs. Contact the faculty with whom you would like to work (see below for the timing of these emails) and ask if they expect to have a postdoc position available around the time that you plan to finish your PhD. You can know how likely they are to have a position beforehand by checking on their funding situation. You can search for their current grants from NIH[3] and NSF.[4] If they don't have one of the larger types of grants, then a postdoc is probably not possible, but it never hurts to ask in case they have some secret money coming from the university, access to a training grant, or some other source of funds.

Some people can work with any kind of mentor. In this case, you can base your interest in a specific mentor solely on their CV and the fit with your interests. However, my experience was that I wanted to work with a fairly hands-off mentor or co-mentors. I had worked enough jobs in construction and foodservice to know that I become temperamental under micromanagement. I needed to get an idea of what I was getting into regarding a mentor's style. As we advised you to do in previous chapters, at conferences, I asked the mentor's current students and postdocs what it was like to work for the mentor. It is always possible that some had a less than great experience due to factors outside the mentor's control, so make sure to ask more than one person if you can.

If you send out several emails to potential mentors and do not have excited responses asking to interview you, do not be deterred or surprised. Having a postdoc-sized amount of money lying around means having about $70,000 US for salary and benefits that you were not planning on using. This is fairly rare unless some other postdoc has recently decided to leave the laboratory, or the mentor has a new grant.

Work with your current mentor, friends, and colleagues to come up with a list of 5–10 potential postdoc mentors to contact. Some names might be from labs you considered

---

2 http://neurojobs.sfn.org/jobs
http://psychjobsearch.wikidot.com/

3 https://report.nih.gov/

4 https://www.nsf.gov/awardsearch/

when you were applying to PhD programs. Start at the top of the list and send them emails. Some of these mentors might be in your department, or your graduate mentor might even be interested in paying you to stick around if you are really kicking butt in the lab. Another possibility that some people pursue, like I did, is to work with more than one mentor. For example, you could split your full-time postdoc position between a lab that does computational modeling and one that does empirical neuroscience, allowing you to essentially do two postdocs at the same time while costing each lab only half of the hiring expenses.

A more important issue than where to find a postdoc is when to look. You should start looking for a postdoc at least a year before you need one. I recommend to my trainees in graduate school that they start talking to potential postdoc mentors at conferences about two years before their projected dissertation defense. This may sound crazy, but it is not. Knowing that you have someone excellent coming to your lab in two years is much easier to plan for than the surprise email from someone looking to start a postdoc in a couple of months. Also, a well-funded lab will be able to bring you out for an interview at any time, well before your start date.

Do not fall victim to the idea that you can send a couple of emails or Skype a couple of people and nail down a postdoc in a couple of months. You really want to start looking for this position at least a year before you need it.

WHAT WILL YOU LEARN, DOROTHY?

Your postdoc years can be a truly magical time. There are no classes to take or to teach. You are not on committees yet. The goal is to spend all your time in the lab, write papers, and do what most of us consider the really fun parts of our jobs. You have enough knowledge to think of your own research questions, design an experiment, analyze the data, and write a viable manuscript reporting the findings. As a postdoc, you can hopefully repeat the process that I just rattled off about 5–10 times so that you are ready to be completely independent by the end of your postdoctoral training, turning out excellent publications along the way.

As you've read elsewhere in this book, the skill that takes almost all scientists the longest to learn and perfect is writing. If things went really well in graduate school, you probably helped write about 10 papers, maybe more, probably less. Note that 10 is a large number for someone doing cognitive psychology or cognitive neuroscience. If you are working with animals, such as nonhuman primates, a large number is more like four, because those methods spend so long in the data collection phase.

Your papers from grad school are a good start. However, in grad school you write those papers with the help of your advisors and mentors who heavily revised your work. If you were really lucky, you had an advisor who explained why they were changing your paper the way they did. But the odds are that you were expected to pick up on the logic of what they were doing without explicit instructions about the principles of scientific writing. A postdoc is the ideal time to practice writing independently in a safe environment where tenure is not yet on the line.

When you take a postdoc at a different institution from where you got your PhD, a big advantage is that the faculty at that institution know a bunch of different stuff from what you learned at your graduate school institution. At your postdoc university, the faculty may have similar, but different research interests from the faculty at your graduate school, and you will learn about tasks, topics, and theories with which you were previously unfamiliar. These can become the focus of your research if you find them particularly interesting.

SKILLS FOR THE FUTURE

There are at least four things that you want to learn during your postdoc: (1) how to write efficiently, (2) the methods you will use during the first 5–10 years in your own laboratory, (3) how to get grants to fund research in your future lab, and (4) what your mentoring style will be.

WRITING

I cannot stress enough how important it is to become a very strong writer during your postdoc. This is the time to write papers from start to finish without getting comments or input from anyone else. You may have co-authors, but you will want to write as though you are going to submit a paper for publication without running it past them. If my colleague Gordon Logan is correct, that you learn to write papers by writing 20 papers, then you probably have a bunch of writing to do to get to 20. You likely have several papers' worth of findings from graduate school to write up. These are great papers to work on while you are collecting data in your new postdoc lab. In addition, you want to establish that you are responsible for your own success by publishing a paper from your new institution as quickly as possible.

As we discussed in the grad school chapter (Chapter 4), the expectation is that you are publishing about three papers per year when you apply for your faculty position. Again, if you are doing animal work, the target number should be more like one paper per year. If you write up work from grad school during the beginning of your postdoc, within a year or so you should have most of that work written up. You should set goals for yourself to write up a paper every 2–3 weeks. This is completely doable. Write for 45 minutes to 1 hour every day. Or set aside a full day or two per week for writing. Once you get good at running your experiments, you can frequently write during the downtime of data collection. Once you get good at writing, you can write the basic text of a paper in several hours. Early on this will likely take two or three times longer, but even that is basically just a couple of weeks' worth of writing.

After getting the stuff from graduate school written up, you can begin writing up your new postdoc projects. To help you think about the structure of papers explicitly, and to boil down several tips and style suggestions that I learned throughout my career, I have made a writing checklist available that we developed in my lab.[5]

LEARNING THE METHODS

The second thing you need to learn during your postdoc are the methods that you plan to use in your own laboratory during the first 5 years or so. Those will be the methods available to you in the first grants that you write because you have to show reviewers that you have experience with the methods proposed in your grants. Because of this, you should think about the tools that will be available in labs you are considering for your postdoc. One strategy is to pick another lab using the same tools as you used in grad school, and then focus on different questions, or use other tweaks. For example, you might go from one fMRI lab to another fMRI lab and focus on a slightly different topic than you did in graduate school. This can make you a true expert on a set of methods. I have seen many people use this strategy successfully to get a faculty position and grant funding fairly quickly.

Another approach is to pick a lab in which you will learn a different set of methods, perhaps in the same theoretical space. For example, you could go from an fMRI lab studying memory to a rodent electrophysiology lab studying memory. This a strategy will broaden your worldview and give you options regarding the type of laboratory that you set up when you get your faculty position. However, this kind of postdoc position will require more time than going into a lab with methods you already know. In particular, if you go from a human cognitive neuroscience lab (e.g., fMRI, electroencephalogram [EEG]) to an animal lab (electrophysiology or optogenetics), you will need 4-6 years to become expert enough with the equipment and animals that you could really set up and run your own lab. This is because some of these methods involve learning surgery and far more equipment than that used in human cognitive neuroscience labs. The time involved in learning a new method during the postdoc years discourages many people. However, in my experience, it resulted in a breadth of knowledge that you cannot gain at any other time in your career. A sabbatical is not enough time to learn an entirely new method or area of research. But a postdoc presents an opportunity to become a kind of scientist that the world has never seen by blending what you have done in graduate school with a new line of work from your postdoc lab.

GETTING GRANTS

The third thing that you need to learn as a postdoc is how to get grants to fund your research. Obviously, the amount of funding and the type of grants you need will vary

---

[5] http://www.psy.vanderbilt.edu/faculty/woodman/CheckList.pdf

depending on your method. Grants with animals or methods like fMRI, require larger amounts of money. However, many of the principles for grant writing are common across methods and funding mechanisms.

In graduate school, you probably saw one or two examples of your mentor's grants. Maybe you wrote a pre-doctoral **National Research Service Award** (or **NRSA**, awarded by NIH) or NSF fellowship grant. However, a postdoc is the time for you to really get practice writing grants and see the grant writing process from start to finish. Obviously, you should write an NRSA (if you are a US citizen, a European Union (EU) grant if you are from Europe, an NSERC, Natural Science and Engineering Research Council, for Canadians, etc.). Establishing a track record of funding your own work is one of the most important, concrete things you can do as a postdoc.

In addition to submitting your own grant, you should become heavily involved in preparing grants with your mentor or mentors. Ask what pilot data you can collect for a future grant. Ask if you can see all the secondary sections (e.g., budget) instead of just the research strategy and specific aims of their funded grants. Ask if you can proofread the grants they are working on. I would be shocked if someone turned down your offer to proofread and help. Learning to write grants is mostly a matter of repetition. Getting as much exposure to grants as possible during your postdoc will help you write strong grants and get funded more quickly once you are a faculty member.

MENTORING STYLE

Fourth and finally, as a postdoc you will need to figure out how you are going to mentor your undergraduates, grad students, research assistants, and postdocs when you have your own lab in a few years.

You had one example of how to be a mentor from your graduate advisor. Now that you are a postdoc, you have one or two more faculty mentors. Generally the major continuum that differentiates mentoring styles is level of oversight. Are you going to meet with your advisees in individual meetings each week? This is the norm, at least for new mentor-mentee relationships. Or are you going to spend a couple of hours in the lab each day and talk to each of your folks a couple of times per week? Or will you occupy the other end of the continuum and work with your trainees mostly electronically, on manuscripts and grants? I believe that most people come in wanting to be hands on, guiding their trainees through weekly individual meetings, perhaps with additional daily conversations throughout the week. What I have seen is that most people migrate toward being more hands off as they mature throughout their careers. There are large differences across individuals with regard to level of oversight.

I started off with weekly meetings because this is what all my mentors had done with me. However, I moved toward just dropping by the lab once a day to talk to everyone who was around. This results in the people who are regularly in the lab getting the most attention. That seems about right to me.

What made me change my mentoring style? What I learned was that my ability to change trainees' behavior in the ways that matter the most is extremely limited. Some people seem to struggle because they do not think about what they read, how it applies to what they are studying, and how what they study is related to daily life. The only suggestion I have is that they allow their mind to chew on sentences and paragraphs as they read. They can pause and ask themselves how what they're reading applies to the experiments they're running, or related to other papers they've recently read. I think this is what everyone does, so it hardly seems like useful advice.

Another common problem seems to be an inability to get the necessary work done. I'm not sure if this is a problem of gumption, moxie, or grit, but I have not found a way to solve it. You can provide feedback that people need to do more and have detailed conversations about time management, but ultimately it is up to the person to follow through, and follow-through seems to happen in a minority of cases after the problem is present. Because of all this, my own mentoring style has become more laissez-faire than I initially envisioned when I started. This may happen to you too, but the important thing is to develop a philosophy as a postdoc because you will be mentoring undergraduates and graduate students as soon as you get a job.

How you are going to mentor your trainees needs to be something that you have in place when you start your faculty job. You will need to describe your mentoring and teaching style in your teaching statement, which will be part of your application materials. Potential trainees are also going to want to know what your style is when you are recruiting them. It is important to have a policy in place before you start so that you do not fumble through these important relationships by trial and error. It can doom your lab to begin with a couple of disgruntled grad students or postdocs who scare away the high-quality prospective graduate students you interview during year 2–4 of your tenure clock. A popular move to avoid this problem is to ask your trainees what they think they need. Tell them the options: weekly individual meetings, dropping by the lab or their office a couple times per week, send their manuscripts back to them as fast as possible and stay out of the way, and so forth. You can always adjust as needed. The most successful trainees were not successful because they were ridden hard by mentors that did not allow them to make decisions and learn from their mistakes. Just remember this fact as you choose your way.

HAVE AN EXPLICIT CONVERSATION WITH YOUR MENTOR ABOUT INDEPENDENCE

What to do on day one? The first thing you should do when you arrive in your postdoc is to meet with your postdoctoral mentor and ask what experiment you should start preparing to run on that day. This will sound familiar to those reading this book in a linear fashion because we also said this about starting graduate school.

Immediately after asking the question about what they envision you running in their lab, ask about the plan for getting you an independent line of work to begin in your own lab when you leave.

It may seem aggressive to ask about an exit strategy on day one, but it is extremely useful to have both short-term goals and long-term goals so you have a roadmap of where you will end up. If you don't know how you plan to exit, you don't know where you are going. My general plan is to start people off with an experiment from an existing grant that they can then spin into their own experiments that they think up later. I like my trainees to think up their own projects, perhaps extensions of their graduate work, that can potentially translate into independent papers, assuming I do not contribute intellectually to the work.

You should also have an explicit conversation with your postdoctoral mentor about productivity. As we discussed in previous chapters, different areas of the field have different productivity standards. It would be useful for you to know what the top candidates in your area look like. How many papers are they publishing a year?

HAVE THAT GRANT READY

A postdoc is a very good time to think about what you want to study in your own lab. In graduate school, I had been trained to think about topics that my graduate advisor thought about. I needed to branch out and develop my own approach and set of questions. To help me think independently, I was given the excellent advice of keeping a notebook with my experiment ideas that were outside the domain of the projects I was working on at the time. But this was not enough to constitute a cohesive program of research that could become a major research grant.

My postdoc allowed me to work through a number of the ideas in my notebook. The papers that grew out of testing those ideas became the backbone of the work that established my independence and helped me get an NSF grant in my first few years as a faculty member. Interestingly, that NSF grant proposed a set of experiments that followed up on my dissertation work, but the postdoctoral projects were vital in making my CV viable for a major research grant.

During my postdoc, I wrote my first **R01**. An R01 is a category of grant given by the National Institutes of Health (NIH) to fund entire labs. That R01 grew out of my training in graduate school and the research I was doing as a postdoc. Ultimately, that grant got funded and has since been renewed several times.

The point of telling you these just-so stories about my major research grants is to encourage you to write your first NSF or NIH grants as a postdoc. Some of you will write a **K99/R00** grant, a great mechanism known as a "Pathway to Independence Award" from NIH, which basically starts you off with an R01 (which is the 3-year independent phase, known as the R00 phase). Even if you do not submit a K99, you should write a grant while you are a postdoc. Take your dissertation topic or whatever topic the majority of your papers are on, add a dash of the methods or perspective that you have picked up as a postdoc, and have the grant written by your second or third year as a postdoc. This way, when you get your faculty job you will simply add some more pilot data from your own lab to demonstrate that you have all the needed equipment, and submit that grant. Your first couple of years as a faculty member will be extremely busy. The postdoc years are the ideal time to lay all the groundwork for grant writing now.

DEFER YOUR START DATE TO THE FOLLOWING YEAR

After a year or two of your postdoc, when you have begun publishing findings from your new institution and have data to present in a job talk, it is a good idea to apply for the best job ads that you find in the fall. Each year you will likely apply for a few more of these jobs so that you end up applying for a larger number year after year. The next chapters will tell you how to get these jobs.

You should always defer a year when negotiating your new faculty job if you're in a postdoc position. This has become standard. It will allow you to collect pilot data for your grant and complete manuscripts for your postdoc while your new institution completes remodeling of your new lab space. Each time my people have not negotiated a deferral, they end up sitting around for 3–12 months before they can even get into their new lab space for a variety of reasons (e.g., construction running behind the promised schedule, delayed equipment ordering and delivery).

SUMMARY

It is easy to look at a postdoc as a necessary evil that you wish you could skip. A postdoc can seem like it's just delaying getting your life started. This is far from the truth. A postdoc is more like a multi-year sabbatical that you are granted after getting your PhD in which you also get a raise. You will learn things during your postdoc that you will not have the time to learn at any other time in your career. You will also be investing in setting yourself up for success in your ultimate faculty position. For many of us, this was the most productive and exciting time of our careers. I hope it will be for you too.

> **Ashleigh's Interview with Elizabeth Buffalo, PhD**

buffalomemorylab.com

Elizabeth Buffalo, PhD is a Professor of Physiology and Biophysics at the University of Washington, Seattle, Washington. Dr. Buffalo is also the chief of the Neuroscience Division of the Washington Primate Research Center. She trained in graduate school at the University of California San Diego (UCSD), working to understand the nature of long-term memory across species and patient populations. She then did a postdoc with Robert Desimone at the NIH. We were interested in hearing Beth's opinion about how to get the most out of a postdoc for several reasons. Primary among them is that her work provides a beautiful example of how to blend graduate and postdoctoral training to form an independent line of work in one's own lab. Numerous funding sources, as well as awards, including the Troland Award from the National Academy of Sciences, have recognized the work that has grown from this innovative combination of perspective and methods.

BETH'S POSTDOCTORAL EXPERIENCE AT NIH

I started by asking Beth why she did a postdoc because I know that it can be difficult to decide to pursue additional training after graduate school. Beth said that she pursued a postdoc because both she and her mentors expected that she would need to go on to a postdoc after graduate school. She said that at that point in time, the additional experience was necessary to get a research job.

Beth decided to pursue a postdoc with Dr. Bob Desimone at NIH to study neurophysiology. She picked him because she was interested in memory, and at the time, he had been studying neurophysiology and memory. This overlap was intended to allow her to learn a different set of skills from what she had learned in graduate school. However, when she arrived at the NIH to work with Bob, he had switched his focus to studying attention, which was initially a surprise because it was so different from what she had been working on in graduate school. Beth said it actually turned out to be wonderful because she was able to learn a whole new field and then when she started her own lab she could go back to studying memory, which she was really passionate about. The advantage to this is that she clearly had her own independent line of research when she started her faculty position. It is important for young faculty to establish their independence from their graduate school and postdoc mentors and Beth was ideally set up for this.

Another unique advantage of her postdoc was that the lab was at NIH and Bob gave her lots of flexibility to pursue her interests and she was able to work and network with a variety of individuals. For example, Beth had the opportunity to work with Dr. Alex Martin to learn fMRI while she was waiting for all the data to come in on her nonhuman primate neurophysiology experiment, which can take a long time. Working with Alex gave her the chance to do a faster experiment, learn a new technique, and get additional publications from that line of work.

Another experience from her postdoc that Beth tries to emulate with her own postdocs is that Bob gave her opportunities to go to small conferences and present work. For example, if Bob could not accept a speaking engagement, he would often suggest a postdoc go in his place. These experiences were critical for getting to know other people in the field. These interactions subsequently served her well when she went on the job market and when she applied for grants.

Before she went on the faculty job market she wasn't sure she could get a faculty position, which felt really daunting. Since she was living in Washington D.C. and working at NIH, she was able to take time to explore other career paths. She had friends who were working in the government, so she contemplated a career in science policy. She had met people who were serving as program officers, so she discussed their positions with them. She thought about being a staff scientist in the NIH intramural program. Beth found all those conversations useful because it confirmed that she really wanted to have her own lab. She found the unique opportunity of thinking through all her career options to be very helpful.

WHAT POSTDOCS BRING TO THE TABLE

I asked Beth what she thinks made her attractive as a postdoc, because the nature of most postdoc positions involves learning a new technique. I wanted to know what Beth felt she brought to the table that made Bob hire her, and why people like Alex Martin agreed to do these side projects with her. First, Beth had one set of skills that were applicable to Bob's lab because she had already worked with nonhuman primates as a graduate student. This was attractive to Bob because when Beth came into the lab she didn't need to be trained on all aspects of the lab's research methods. Second, Bob talked to Beth's graduate school mentors and he seemed to have largely made the decision to bring her on based on their recommendation. Third, for the fMRI experiments, Beth brought time, enthusiasm, and independent ideas. She had some detailed anatomical questions and hypotheses based on what she studied in graduate school. She also brought expertise with the stimuli and task, which she had developed in graduate school. These aspects of interest and independence meant that she was valuable to Alex, who had scan time to offer and the ability to mentor her through the project.

WHAT BETH IS LOOKING FOR IN A POSTDOC

Beth told me that when she first started her lab she was apprehensive to take on postdocs. Although she felt confident that she could help secure a postdoc for her graduate students, she felt less confident in her ability to help a postdoc get a really competitive research position given that she had only just gone through the process herself. However, Beth's confidence in mentoring postdocs has grown. One of the factors that she said has helped was her experience serving on faculty search committees and grant review committees. This enabled her to better mentor her postdocs on how to write a good grant, prepare job application materials, and give a good job talk.

When interviewing postdocs, Beth thinks about how the postdoc appointment might help them in their career path, leaving open the possibility that their career path might not be a faculty position at a big research university. She spends a lot of time during the interview talking about their goals. If the goal is not to have a faculty position, then she may still take them on as a postdoc provided she can tailor the experience to their goal, and as long as they want to work on Beth's funded projects. In making hiring decisions Beth says it's important to figure out a reason why the postdoc position would be a useful experience for the potential postdoc. She has developed this strategy over the last few years. This represents a shift from her initial approach of thinking she needed to get all postdocs a job at an R1 university. Rather, she thinks she wants to support whatever their career goals are and only about half her postdoc applicants are interested in R1 jobs. The remaining half are interested in industry (e.g., big data, data science), teaching, and staff scientist positions.

I loved how open Beth was to supporting her postdocs' interests because so many faculty seek to train mini versions of themselves, replicating their own career in their trainees. This is most natural since it is the experience that we, as faculty, have available. However, Beth said, "I think it's great to explore other options that are going to fulfill your life and let you do science in the way you want."

I explicitly asked Beth what she weighs most heavily when she decides to take on a postdoc. The first thing Beth said is that she cares about what they have done. Were they productive during graduate school? Then she heavily weighs the recommendation of the graduate mentor. She doesn't require that she knows the recommender, but that the graduate mentor indicates that the applicant is enthusiastic and has a passion for the work. Beth is much more concerned with the energy, enthusiasm, passion, and record of productivity than, for example, the name brand of the school the applicant comes from.

I asked Beth whether she was concerned about applicants justifying why they wanted to pursue a postdoc. In Beth's area of animal research, virtually no one gets a job right out of graduate school, so the applicants do not have to explicitly justify why they want a postdoc. The only red flag might be if they are looking for a second or third postdoc and they do not seem to be looking to acquire new skills in each new postdoctoral position.

OPPORTUNITIES FOR TRAINEES

Beth is part of the Simons collaboration,[1] which has special meetings where postdocs can present. These are great experiences to get ready for job talks, to network, and to promote her postdoc's work. Beth also encourages them to teach summer science courses to help with teaching skills and to network at local institutions. Similarly, Beth invites postdocs to give lectures in undergraduate courses. She said that experience lecturing actually helps them network at their own institution, and that is important to develop such relationships for letters of recommendation. When they are applying for positions they will need at least three letters, so if Beth is one and their graduate mentor is another, the postdoc still needs to be networking to get a third letter from someone else.

Beth often has postdocs co-author review papers. She finds this necessary to help build her postdocs' CVs because nonhuman primate work can take a long time to publish. Writing these reviews also gives the postdocs a chance to develop their own voice. Beth has found that writing review papers has helped postdocs develop their own ideas for experiments, grants, and even entire lines of work to pursue when they get their own lab.

Depending on the trainee, Beth's postdocs all write grant applications, either NRSA grant applications or K99 applications, if they are eligible for them. Beth has some postdocs who have gotten involved in writing her R01s, which gives them experience writing grants and being involved in all aspects of grant preparation. She echoed Geoff's sentiment to make sure postdocs read all the non-science parts of grants to learn all the pieces that go into a grant (e.g., thinking through the budget, understanding how to describe facilities). Beth likes this experience of helping the postdocs learn grantspersonship prior to setting up their own lab.

THE BENEFITS OF HAVING A POSTDOC IN THE LAB

I asked Beth about the structure of her lab and how postdocs fit into it. She said that as her lab has grown, teams of people have become dedicated to working on specific projects, meaning that there are numerous multipersonnel projects, which might also include collaborations with other labs. Therefore, along with her

---

1 https://www.simonsfoundation.org, https://www.simonsfoundation.org/team/elizabeth-a-buffalo/

weekly lab meeting, there are individual team meetings several times a month. One important role that postdocs play is leading those team meetings with the grad students and technicians when Beth is unavailable. This helps postdocs learn how to organize and oversee a team, and learn different managing strategies for different people, in preparation for employment down the road.

Beth says that of course there are many other tasks that postdocs are uniquely suited to do, in addition to simply delegating tasks in the lab. Although Beth has had exceptional graduate students, she said that postdocs have the potential to have deeper discussions about the research because they know the literature better than graduate students. Postdocs are often more interested in developing their own voice through writing grants and manuscripts and giving oral presentations. Postdocs often have their own experimental ideas that can lead a lab to new and interesting questions. Postdocs also have more time to think deeply about their work relative to graduate students, especially since graduate students in their first few years juggle courses and teaching assistant responsibilities. Beth says that another advantage of postdocs is that they bring a whole set of expertise that they developed as a grad student (e.g., they've read different papers than Beth has, they've had different advice from their graduate mentors, different ways of thinking), and this expertise provides an entirely new perspective and voice to lab discussions.

EXPECTATIONS OF INDEPENDENCE

Beth said that she individually tailors her expectations for independence for her different trainees. One important reason for this is because it depends on the funds that the postdoc is being hired with. For example, if a project is funded and ready to get started, and the grant funding for that project is the money that will pay the new postdoc, then the postdoc needs to be excited about being plugged right into that project and hitting the ground running. However, if general institutional money is being used to hire a postdoc, there is more of a possibility to pursue a project the postdoc is interested in outside the aims of the lab. Beth said she certainly supports postdocs in developing their own separate exploratory question on the side, which may lead to its own line of work, while they are working on grant-funded research.

In terms of helping postdocs establish independence that will help them on the job market, Beth has found that (1) inviting students to join her in writing a review article, as previously mentioned, (2) giving them time to think, and (3) helping them work through their job application materials helps postdoc trainees the most. Through this process, Beth helps postdocs think about how to create their own line of work that will be independent from hers.

TIPS FROM BETH

- Beth has seen a lot of people advertising on Twitter for postdocs lately. If you are on Twitter and follow scientists in your field who are really active on Twitter, often they will re-Tweet postdoc ads.
- One way that Beth helps her graduate students get postdocs is to send her own email to the faculty members when her grad students reach out to faculty inquiring about potential postdoc positions. Everyone gets a ton of these emails from postdocs and Beth finds that sending her own email helps her students' emails get noticed. In her emails she writes that she'd be happy to answer any questions or give a formal recommendation to increase the odds that the inquiry gets noticed.
- Beth said that to be competitive for a postdoc in a nonhuman primate lab, the unwritten requirement is that grad students have at least one paper submitted, hopefully published, by the time they inquire about a postdoc. So for different fields the expectation for how frequently students should be publishing would also be different than what Geoff outlined above.
- Some institutions have professional development courses for postdocs that can be a kind of boot camp for setting up your own lab. This may include presentations about hiring, negotiating, grant writing, et cetera. At Beth's institution there is also a postdoc seminar, giving senior postdocs a chance to perfect their job talk skills for a wider audience. Inquire about these opportunities at institutions as you are considering postdocs.
- Beth tries to be supportive of her postdocs taking time off because a postdoc is a difficult position in which there isn't clear time off. You might always be at a critical point in an experiment which can lead you to believe there is never a good time to get away. If a postdoc approaches her with a desire to travel for a week, Beth is really interested in working with them to find a good time to do it. She said that such situations are when it is really critical to have a good team mentality. Ideally, someone else can pick up the slack and you can return the favor down the road.

CHAPTER 6

How to Get a Job at a Teaching School

by Ashleigh

At some point during your graduate school career, you may decide that you plan to apply for faculty positions that weight teaching more heavily than research. You might make this decision because you have a passion for sharing your subject matter through teaching, you thrive on the human interaction that comes with interacting with classrooms full of students on a regular basis, or you do not want to measure the success of your life's work by impact factor or grant funds. You may be steered away from this decision by well-meaning mentors if you are at a research-intensive institution. This can happen for many reasons, including that your mentor is better prepared to advise you on a career path that is similar to her own, because your mentor views success as having a research-heavy career and therefore that's what they want for you, or it can happen because of your department's desire to improve its rankings if you land a research-focused faculty position.

WHAT IS A TEACHING JOB?

In this chapter I talk about the complicated, unspoken rules that will help you land a so-called teaching job. Frequently, a *teaching job* refers to a job at a **small liberal arts college** (often referred to as **SLAC**), which are also typically private, meaning that these institutions are not operated by state governments. Small liberal arts colleges primarily specialize in educating undergraduate students in liberal arts and sciences. Historically, many liberal arts colleges were founded to offer instruction for careers outside those offered by government-run state universities, which frequently specialized in fields like agriculture or dentistry. Instead, small liberal arts schools typically offered degrees in fields such as education and philosophy. A liberal arts education is one that emphasizes developing well-rounded students, often by requiring them to take a broad variety of courses, conduct service in the community, and attend lectures on a variety of topics throughout their undergraduate career. It is key to these institutions that their faculty foster an environment of nurturing support for their students. Although there are other institutions with teaching jobs, such as associated faculty positions at state schools or research-intensive schools that have distinct positions for research professors and instructors, here I focus on liberal arts jobs as the model for teaching jobs.

Many institutions categorized as small liberal arts colleges have high teaching loads for their faculty, but it's important to acknowledge that some have very intense research expectations and I am not discussing those schools in this chapter (see Chapter 7 for advice on landing research-focused faculty jobs). Some state schools also require a teaching load similar to that of teaching-heavy liberal arts colleges, accompanied by relatively low research expectations. I worked at a state school that had the same teaching load (24 credit hours per year) as my previous employer, a small liberal arts school. Therefore, here I am focused on providing advice for securing a faculty position at any institution that has a relatively high teaching load, regardless of its status as either a liberal arts, private, or state university. Getting a teaching-heavy job is a different ballgame from getting a research-heavy job. To understand how to land these jobs, it's important to understand why the characteristics of the ideal candidate are not immediately clear and how you can better groom yourself to be well positioned for a faculty job at a teaching-heavy institution. Throughout this chapter I will use the terms liberal arts school and teaching institution interchangeably. That does not indicate that there is no difference, but rather than this advice applies broadly to both environments.

It can be hard to navigate getting a tenure-track faculty position at a liberal arts institution. This is because the expectations placed on faculty of liberal arts colleges across the country are changing. Facing the ever-demanding pressure to illustrate the benefit of spending tuition dollars on a liberal arts education in a strained economy, departments at many of these colleges measure and advertise their success based on employment rates of their graduates. For many fields of study at liberal arts institutions, however, their measure of success is based largely on the proportion of students accepted into graduate programs. For example graduate school is required for psychology majors who want to actually be psychologists; you generally cannot be a licensed psychologist with only a bachelor's degree. In order to gain admittance into competitive graduate programs, undergraduate students need high quality research experiences. This means new faculty at liberal arts colleges are expected to have both teaching and research expertise. While this expertise benefits undergraduate students immensely, it does create an infrequently discussed expectation that is difficult to navigate for the average faculty applicant.

Typical tenure-track positions at liberal arts schools include a 3–4 class (12 credit hour) teaching load each semester. Faculty must also maintain an active research program to provide research opportunities for students. The required gravitas of this scholarly activity varies widely across institutions. At many institutions, it is preferred that the faculty stay active in their field, publishing, peer reviewing, and attending conferences so that the letters of recommendation the faculty write for their students carry more weight.

However, at other institutions it is sufficient to present research projects at a regional undergraduate conference. I will first talk about why these expectations are unclear and then turn to how to better understand them. This relatively high expectation of research productivity, given that it co-exists with a high expectation of teaching ability, is not clear to the average applicant for at least four reasons.

First, jobs at these schools have often been called teaching jobs, implying that teaching is all the faculty are expected to do to earn tenure. The term *teaching job* is misleading because if all faculty at these institutions did was teach their assigned courses, students wouldn't succeed, and without successful alumni to donate to the institutions, and alumni's success stories to attract new students, the teaching institutions would likely be in real financial trouble. Faculty engage in many activities outside the classroom that help promote their students and institution and inform their teaching, such as publish research, bring in grant money, network at conferences, serve on committees, peer review, and work with clients in private practice.

Second, the seasoned faculty members who have been at these institutions for many years were not hired under the current high demand for research productivity and may not have active research programs. Thus when the applicant searches for the scholarly activity of the current faculty, they are likely misled in terms of current expectations for faculty research activity. Specifically, if you look at the publication record of folks who were hired 20 or 30 years ago, they may not have published at all in the past decade. Don't be misled by this and think that you can also get the job (and ultimately tenure and promotion) without having an active research program. Unfortunately, looking at the profile of current faculty members in your targeted department is usually a poor way to judge your fit for that institution. In the section below I will provide some solutions to judging your fit as an applicant when looking at faculty profiles is not a fruitful method to determine your likelihood of success at a school.

Third, the established tenure requirements, if transparent, generally include very low expectations in terms of research. Often, it's technically possible to get tenure at these schools with few or no publications. These advertised tenure expectations, however, don't reflect what the search committee is actually looking for when reviewing applications. One reason for this is because with the competitive job market, the search committee can find and secure a better candidate than someone with little to no promise of a generative research program. Therefore, even though it may technically be possible to secure tenure and promotion at the targeted institution with little research experience, the search committee likely has plenty of applicants who have reasonably successful research programs who rise to the top of their applicant pool. A second reason the published tenure expectations might be more lenient or open-ended than what you ultimately experience if hired there, is that they are often written to broadly apply across a variety of disciplines. Tenure outlines need to accommodate some fields that don't use published research as their gauge for success. In order to overcome the misunderstanding that can be presented by requirements written for the whole college, be sure to outright ask about them.

Fourth, the liberal arts are about much more than you can measure by a line on a CV. Many faculty members at liberal arts institutions understand that getting students jobs is not their only goal. At these schools there are expectations for interdisciplinary work, integrating ideas across the curriculum, encouraging students to exhibit a breadth of knowledge, and applying their knowledge outside the classroom in service to the community. In our chapter on getting tenure at a liberal arts school, I interview Glenn Sharfman, who echoes the sentiments that much of what makes these schools unique may not be measured as a line on a CV. Looking up the current faculty may not illustrate what the individual faculty members have done to contribute to the institution. Therefore, it's a mistake to judge your fit with an institution based on first appearances.

The astute applicant for these positions is often frustrated that they need to be both excellent at teaching and research to get these jobs. Indeed, the PhD is a research degree, and in grad school there is typically little to no focus on helping graduate students hone teaching expertise. Although this may seem frustrating, it's unrealistic to expect a PhD program to provide teaching experience. However, with the help of some good advice (look at you, holding this book with good advice flowing from its pages!), there are some simple adjustments graduate students can make while they are earning their PhD to better position themselves for faculty positions at liberal arts institutions. I outline these suggestions below.

HOW TO PREPARE FOR A TEACHING JOB

GETTING TEACHING EXPERIENCE

The old adage that you can't get a job without experience, but you can't get experience without a job also applies to liberal arts teaching jobs. Unfortunately, it's hard to get a teaching-focused job without teaching experience. However, because the PhD is a research degree, it's unlikely that most graduate students who are gathering materials for their job applications have substantial teaching experience. Here, I suggest a few creative ways to gain some form of experience that can at least pass for teaching-related experience in your application materials, likely surpassing many other applicants who did not receive this advice.

In terms of teaching experience, graduate students can gladly accept teaching assistantships. Although often frowned upon by research mentors because they take time away from research, teaching assistantships do provide an insider's view on what it takes to run a class. Further, many institutions offer some type of teaching certificate program; listing this on your CV as teaching experience is certainly better than no teaching experience at all. Finally, if you are certain you are going to apply for teaching-heavy faculty positions, it might be wise to get a Master's degree while still in graduate school, if your program offers that option. This would allow you to apply for local adjunct positions. Many community colleges only require a Master's degree in order to adjunct for undergraduate classes, so it may be possible to teach a course at a local college while still working on your PhD. This would allow you to have your own course, demonstrating that you've managed all aspects of teaching, not just those allocated to you as a TA. This is only a reasonable option if it does not significantly interfere with research productivity; after all, you need to successfully complete your PhD to get a faculty job. It's true that there is always research-related activity you could be engaged in, from writing grants to reviewing papers to reading journal articles, but if you are going to apply for jobs that value teaching experience, you'll need to be thoughtful about managing your time to best set yourself on track for those positions.

At a number of PhD-granting institutions, like the graduate program at Ohio State, there are great opportunities for graduate students to teach their own sections of Introduction to Psychology. The general model involves a faculty coordinator who essentially spends the summer teaching graduate students how to teach, with workshops on developing their own PowerPoint presentations, learning the course management software, devising assignments and grading rubrics, tying the course into the general education requirements and the freshman common book (i.e., many institutions ask all incoming freshmen to read a specific book that changes each year and then some faculty who work with freshmen are asked to work it into their courses), and other useful skills for the classroom. Then the faculty coordinator supports the graduate students during their first semesters teaching with events and speakers, with constant contact and advice, by evaluating their teaching, and so much more. I am really impressed by this model of preparing graduate students to teach. Although most students and faculty likely don't see the benefit of this system, all it would take is sitting on a search committee for faculty at a teaching school to see its inherent value. If your university offers any type of teaching training program, and you can take advantage of it without jeopardizing your research, do it!

Many applications will require some form of evidence of teaching effectiveness. There is a lot of debate about what truly demonstrates effective teaching. As you establish your teaching experience, it would be wise to work on compiling an online teaching portfolio. In this portfolio you can accumulate and save evidence of your teaching from descriptions of interesting assignments you created for students, student evaluations of instruction, and even video recordings of you lecturing. Many institutions have resources for teachers that include having someone come and record your lecture to provide feedback on your instruction. This is another way to begin growing a teaching portfolio in case you need it for an application down the road.

Some graduate programs even require that students gain teaching experience as part of their program, so you want to keep in mind that when you apply for jobs, you will be competing against applicants with teaching experience. Some graduate programs include an online course that teaches students about pedagogical issues, a teaching seminar, or opportunities to guest lecture, after which you can get written feedback from the students to possibly submit as teaching evaluations with your job application. Depending on your program, you may need to be more proactive than folks at other schools to be competitive.

MENTORING STUDENTS IN THE LAB

One skill you will need to perfect in a faculty position at a liberal arts school is blending teaching and research responsibilities. To get your feet wet in this kind of multitasking, I suggest mentoring undergraduate research assistants as a form of teaching. Graduate students should ask their mentors if they can be in charge of the undergraduate research assistants working in the lab. This benefits everyone involved by lowering the faculty member's workload, giving the undergraduate research assistant more attention, and giving graduate students experience mentoring students. In this mentorship role, you can encourage research assistants to apply for regional conferences and small grants from honors societies to fund research and travel to conferences. You can then add the research assistant's accomplishments to your CV in a section about mentoring students. Depending on the size of the lab and how the faculty member likes to run their lab, it might be ideal for the graduate student to run a separate lab meeting for undergraduates only. This would have the added benefit of freeing up the regular lab meeting led by your mentor for graduate student business. This scenario could demonstrate success with mentoring undergraduates without actually teaching a class. Even if taking over all the undergraduates in the lab is not feasible, mentoring one or two in succession is still impressive on a CV.

TEACHING POSTDOC

At a talk I gave to graduate students and postdocs about how to get a job at a liberal arts school (the talk that later turned into this chapter), I was told that the best piece of advice I gave was that when applying for faculty positions after graduate school, to keep in mind that a career path is infrequently linear. Many graduates will need to consider spending a few years as a lecturer or visiting assistant professor on the way to securing a tenure-track job at a liberal arts school. These years can be considered a teaching postdoc, similar to a research postdoc. I use this term informally, but there really are formal teaching postdoc positions, which we will discuss later, in the interview with Lauren Hecht. I know the idea of moving for jobs multiple times might be an unwelcome concept, especially for people who have families or significant others. However, I offer this suggestion because I know it's daunting to be a graduate student and compare yourself to the faculty members in your department. Their positions often feel incredibly out of reach but it's vital to keep in mind that the first job out of grad school is unlikely to be the dream job, and that's ok. Geoff left graduate school and has been at Vanderbilt ever since. In contrast, I left graduate school and have changed institutions four times. But both of us are really satisfied with our career paths for different reasons. You might love where you end up right out of the gate or you might outgrow your position. Have some confidence, and keep your options open, and despite the permanent feeling offered by the tenure system, don't feel tied down to the first place you get a job. For more suggestions on getting through some years of uncertainty, on one-year contracts, and on potentially living apart from your significant other, see Chapter 12 on work-life balance.

PREPARING YOUR APPLICATION

INSTITUTION-SPECIFIC RESEARCH PLANS

I started this chapter by describing the need to provide undergraduates at liberal arts colleges with research experiences, which in turn helps them gain admittance into graduate programs, which in turn helps the institution's reputation. Given that emphasis, there are some important tips to help you get interviews for faculty positions at liberal arts colleges. First, it's important to contact the chair of the search committee and inquire about startup funds (i.e., the money the school gives you to get your lab going. Startup funds are often spent on equipment and personnel.) before submitting any application materials. Startup costs can vary widely, and research needs must be tailored to fit what the school can provide. Some chairs may not know an exact number, but they will be able to tell you what they have seen before (e.g., an eye tracker) and what is never going to happen (e.g., an fMRI scanner). Second, it's important to have a clear plan on how to incorporate undergraduate students into your research, such as having them run subjects, program experiments, and analyze data. Be sure that these plans fit with what the institution can offer. For example, don't propose that knowledge of a particular software will be a requirement for working in your lab if the institution does not offer a class on it. Also, don't suggest you are going to offer course credit for working in your lab if the targeted department does not have such a system in place. Third, having plans to take the undergraduate students to conferences and apply for funding to take the students to conferences from reasonable granting agencies would be very impressive. Know your audience though: if the conference is not regional or not located in the same town as the school and your students are not upper middle class, your plan to take them to a conference might sound more like an unrealistic dream of an out-of-touch applicant, rather than an engaged, thoughtful applicant. Fourth, a plan on how to apply for grant money to fund additional lab equipment beyond what startup funds cover is smart, especially if you've already selected a specific grant to apply for and mentioned in your application materials. As I discuss below, be particularly mindful of grants that you'd qualify for at this particular institution (i.e., the institution may be particularly eligible or ineligible for certain grants based on things like its size, resources, or its student population). Fifth, examine the school's curriculum and search for an honors program. Working with students through honors or thesis programs can be a great way to work with top students, and you need to know which majors have them so that your proposal can fit nicely with the school's current offerings.

COVER LETTER

At the talk I mentioned earlier that was the starting point for this chapter, the most surprising piece of advice according to the R1 faculty in attendance, in terms of application materials for liberal arts jobs, was the degree of care with which the cover letter should be written because often the search committee is not made up of faculty who share an expertise with the applicant, and therefore they cannot easily skim an applicant's CV and know if the applicant is publishing in top-tier journals and the reputations of their co-authors and mentors. This means that the cover letter is much more important when applying to liberal arts jobs than R1 jobs. As such, cover letters should be on official letterhead to make them stand out (you would be shocked at how few cover letters are submitted on official letter head). Cover letters should include a list of all the classes mentioned in the ad that the applicant may have to teach. You should clearly state the classes you have experience teaching and those you look forward to teaching (all of those that you have no experience teaching). The cover letter should include your experience with student mentorship in research, as mentioned above. If possible, mention a personal connection with the school, even if it is simply an awareness of the school's strong reputation. Always include a desire to live in the specific geographical area of the school if there is one, such as having family in that location or wanting to live in a small college town or community. The cover letter should also include mention of any noteworthy fits between the applicant and the institution's mission statement. Finally, include buzz words from their website that illustrate aspects of the student experience at the institution that excite you, such as student-centered philosophy, service learning, immersion learning, experiential learning, or interdisciplinary connections.

TEACHING STATEMENT

Is it possible to over-state the importance of your teaching statement in applying for a teaching-heavy faculty position? I don't think so, but I also think there are some easy ways to tackle this seemingly stressful task. In terms of the teaching statement, remember that many faculty on the committee were likely not hired with high research expectations and are often very good teachers. Even with the inclusion of high research expectations in most searches, teaching is still considered the primary role of many faculty at most liberal arts institutions. There is some debate over the utility of asking for teaching statements from job applicants who are likely novice teachers themselves (think about having a kindergartener explain their political views), but there are a number of ways applicants with little to no teaching experience can successfully develop a teaching philosophy.

First, ask everyone you know if you can see their teaching statement, ranging from your research mentors, to your professors, to your friends who are on the job market. Keep in mind that you will see a variety of teaching-focused and research-focused application materials and you will likely not want to model yours on the faculty at the institution you are attending if it's an R1 school. However, an example of what not to do is just as useful as an example of what your teaching statement should resemble.

Researching "novel pedagogy at the university level" may yield some interesting ideas, such as the *flipped classroom*, where lectures are all online and class time is used for more interactive activities. Reading about popular classroom techniques you might implement in your classroom in places like the *Chronicle of Higher Education*[1] can allow you to plan out your first semester classroom even with no teaching experience. Check out recent issues of pedagogical journals such as *Teaching of Psychology* that include articles that comprehensively cover hot topics, like staying up-to-date on research on best teaching practices, as described in Chew et al.'s "Improving teaching and learning in psychology" (2018).[2]

Discussing what you plan to do in your first semester of teaching is a great alternative to discussing what you've actually done if you don't have any teaching experience. Really, even those of us who are seasoned instructors often discuss our failures and plans for the future as much as we discuss what went smoothly in the classroom. Demonstrating effectiveness with a technique is always preferred, but if you don't have the experience, at least try to do more than just use buzz words; use specific examples to bolster the type of pedagogy you would like to use in the classroom.

Make sure you personalize your statement for the school you are applying to by using information you find on the university and department's website. You want to make sure the classroom management and assignments you describe are reasonable given the typical classroom size and TA (or course assistant, CA, as they are called at Ohio State) availability. For example, you do not want to say you plan to assign writing assignments, such as reflection papers, in every class period if your course size is likely to be over 100 students. As previously mentioned, you will also want to emphasize any hot teaching methods championed by the institution such as learning communities.

Form a group of folks who are on the job market at the same time as you and read each other's statements over and over and over again. Talking with other applicants is also really helpful when it comes time to get feedback on preparing for interviews. I

---

1   https://www.chronicle.com
2   Chew, S. L., Halonen, J. S., McCarthy, M. A., Gurung, R. A. R., Beers, M. J., McEntarffer, R., & Landrum, R. E. (2018). Practice what we teach: Improving teaching and learning in psychology. *Teaching of Psychology, 45*(3), 239–245.

actually interviewed at the same school as a friend from graduate school and neither of us knew it—that's embarrassing and ridiculous—be talking to your peers who are on the job market. They are your best resource.

Remember that there is often a member of the search committee from outside the department. This person likely understands very little about your research accomplishments and will likely take a keen interest in your teaching statement. It's a good idea to have a theme throughout your statement because it demonstrates that you can present information with a unifying thread. In this way, your teaching statement is a chance to show how coherently you could teach a class. I have a friend who had a clever theme of *hunger* for learning and knowledge who began her teaching statement with a quote by Julia Child and it worked! She got her dream job. Even though you're writing with a theme, be sure to maintain authenticity because you will likely be asked about it during your interview. I wrote about taking a class called "How to be a Great Professor" in my personal statement and the first thing someone asked me during an interview was "Tell me how to be a great professor," and it caught me off guard! You should be prepared to respond to questions about anything you've written, so be true to yourself.

If you have never taught before, you can discuss mentoring students in your lab, attending STEM (science, technology, engineering, and math) teaching discussions or pedagogy workshops on campus, and contemplating other teaching-related issues like academic dishonesty. If there is no pedagogy discussion group, start one. I spoke to a STEM pedagogy group composed of postdocs from a variety of campus departments and discovered that many folks had similar concerns about teaching-heavy jobs, especially because they had pursued postdocs and had strong research records, but little teaching experience. These types of group teaching discussions are a great place to get advice and camaraderie. Further, if you start one of these groups you can invite faculty from across campus to speak at your meetings, which is great for networking.

If your teaching experience is with age groups that are not undergraduate students, you can discuss topics that teaching all ages requires, such as quickly assessing what your students do and do not know so you can meet them in a place where they can learn from you. Perhaps suggest some activities that will help you do this, such as a pretest on the first day of class (e.g., statistics textbooks often have a math pretest in the appendix). You can also describe assignments that will help elevate students to a level at which you can instruct the material if they appear below baseline.

Don't reiterate weak points from your CV. If your only teaching experience is serving as a TA six years ago, talk about it only vaguely in your teaching statement. Do not remind the reader again that you have little experience. One teaching statement I read talked so much about how the person would manage the classroom at the institution where they were applying that I could not figure out from the statement alone whether the author of the teaching statement actually had any teaching experience. I also didn't really care if they had actually taught before by the end of the statement because they clearly had a pretty good plan in place for teaching when they arrived on campus.

When you get to the in-person or phone interview, you could mention that you will ask for syllabi from other courses to implement tried-and-true policies at the specific institution. For example, if students trickling into class late is a chronic issue at an institution, the syllabi will likely have policies that state that faculty will not repeat themselves throughout class to the detriment of other students. Or if classes are very large, dropping one exam might help alleviate the huge burden on the professor of responding to and verifying every excuse students give throughout the semester that might impact their attendance and performance. Reading other syllabi will also help you know what support systems are in place (e.g., a student advocacy center that verifies students' medical and other absence excuses for faculty) and communicate to the search committee that you respect and appreciate their years of experience and hope to learn from them. Despite years of teaching experience, I have discovered that the best way to smoothly transition into a new institution is by mimicking the policies on syllabi in place when I arrive. I have admittedly not understood why some policies were necessary until the semester started and I discovered something that seemed like a trivial policy was a game changer in helping the semester run smoothly. If you say that you are going to do this in your written materials, it might sound like you plan on just copying the other faculty, so this is a piece of advice that might come across better orally in an interview.

DIVERSITY STATEMENT

If you are seeking to work with a population other than your own or are required to write a diversity statement, here are a few tips based on my experience as a white woman teaching at a historically black university (often referred to as an **HBCU**, which stands for **historically black colleges and universities**).

Make it very clear that you are open and interested in learning about the culture and lives of any population different from your own. Asking students questions about themselves demonstrates that you care without jeopardizing your authority in the classroom. While teaching at Tennessee State University (TSU), an HBCU, I asked my students questions ranging from whether they preferred to be called Black versus African American (and why) to whether I could wear a shirt that said *TSU is my HBCU* sold at the school bookstore. These questions led to great conversations that enriched the classroom environment and my bond with the students. One of these conversations led the students to ask me if I preferred the term Caucasian or White, which really blew me away since I had never been asked that question, nor had I ever considered the answer. Another casual conversation led to a student asking me if I had ever taken the beauty test. I told her I didn't know what she meant, and the student entered *beauty* in Google Images and all the images that came up were of White women. I learned some really important lessons about who my students were and what they dealt with on a regular basis that connected me and my students in really powerful and transformative ways. These are examples of how you can learn from your students, and rather than undermine your authority in the classroom, it actually fosters an environment of caring and mutual interest that supports learning.

Although this is really first day advice rather than interview advice, it's worth saying in this section on diversity that your syllabus policies should be sensitive to the population. For example, if you are at a school where students are largely commuters, it's not uncommon for students to arrive at varied times due to traffic. If you are at a school with a largely financially disadvantaged population, be conservative in the price of required course materials. If you have a good number of nontraditional or working students, have reasonable policies in place for when their childcare falls through or they have to pick up an extra shift at work. It's understandable that since you have your PhD, you are probably somewhat out of touch with what the average student deals with on a daily basis. Don't think about what you could accomplish when you were in their shoes—you were above average or you wouldn't have been in a PhD program—be reasonable based on the rigor expected at your institution and be compassionate.

Unlike the suggestion above that is really first day advice, a way of enhancing your teaching statement might be to propose an in-class activity to ask why the students chose that particular school. For example, when I did this at an HBCU I discovered that students chose the school to avoid racism on campus, to learn about their culture, because they had a family legacy of attending an HBCU, or because they had grown up surrounded by people who did not look like them and wanted to have the experience of fitting in during college. This was one of my favorite assignments that really armed me with an understanding of why we need HBCUs. My only regret was not doing it on my very first day on that campus.

RESEARCH STATEMENT

Earlier in the chapter I mentioned some institution-specific issues you need to be aware of in terms of your research program. Here I elaborate on these points as they relate to writing your research statement. First, be sure to discuss how you can involve undergraduates in your research. Make it very clear that you understand that you likely won't have graduate students and that involving undergraduates is meant to serve as experiential learning for the students and not just free labor for you.

Second, clearly communicate that you can set up a lab within the allotted budget. This is where having inquired with the search committee chair regarding the budget for startup costs will prove useful. You will of course need to tailor your needs and plans for your research program to fit within this number. Another aspect to understand is how much data collection is possible. If you require expensive materials or costly equipment for the research you have been doing in graduate school, you may need to consider if there is a method that you could pick up that is not expensive but could still result in publishable research. Most scientists have a research portfolio of projects that require more and less expensive scientific methods. You might need to focus on the lower end of those requirements in your research statement. Finally, offer some ideas about how to secure external funding for additional (but not immediately necessary) equipment.

Third, it's imperative to demonstrate that you can function independently as a research scientist. Although this is true in all positions, the concern at small liberal arts schools is that there is not going to be anyone on the faculty who conducts research remotely similar to yours. This is isolating because you will not have peers to brainstorm ideas with or have joint lab meetings. In addition to establishing your independence, it is attractive for smaller liberal arts schools to have connections with other research-focused institutions, provided that you are clear that these connections are secondary and you will function primarily at your home institution. Thus you should describe plans to continue or establish collaborations with faculty from other institutions to show you bring connections and other ways of being productive. Many conferences and professional societies offer networking grants where they will help fund your travel to work with someone at another institution. However, make it very clear that undergrads from the institution to which you are applying will also be involved in, and integral to, your research program.

Fourth, do your homework on the types of grants that you would be eligible for at this particular institution. For example, there are specific grants for primarily minority serving institutions and others for liberal arts schools. If you have a grant in mind that you intend to apply for, or even a grant application in the works, you will be head and shoulders above your competition and very appealing to the administration. Be prepared to highlight this progress in your job talk or during your interview.

INTERVIEWING

If you get a phone interview, you will likely be told a considerable amount of information about the position at that time. However, this is a chance for you to also ask questions that will help you perform better during your on-campus interview, should you get one. The following tips and questions can be used for the phone or campus interview.

Show off your homework. Know the classes they want covered (they were listed in the ad), know their mission statement, know their student body, know the faculty on the search committee (email and ask ahead of time who is on the search committee if you need to), and ideally, know of all the faculty in the department.

Have student-centered goals with a reasonable timeline for implementing them. For example, have a regional conference you would like to take students to and know a funding source where they might get travel funds to support them (e.g., the honors society in your field). Know the deadlines for both the conference and the funding sources. Mention that you would like to reach out to some strong students over the summer to offer them a research position in your lab and that you would have them submit an abstract to the conference after working in the lab for a year. After that you plan to recruit students who are in your classes, but you would need some suggestions for your first semester of who to recruit. Be sure to also inquire about what you will be able to offer these students. For example, does the department have a mechanism through which students can earn credit while working in your lab?

Be sure to ask about the tenure process; it's concerning when applicants don't inquire about the tenure process during an interview for a tenure-track position. Ask about how transparent the whole process is, ask what the publishing requirements are, and ask how the landscape of research expectations is evolving at that particular institution. You should make it clear that you will go above and beyond the publishing requirements because you understand it benefits the students if you are active in research.

Come with new ideas but present them tactfully. Some possible ideas are below. Some institutions would balk at some of these suggestions (e.g., not all colleges are interested in becoming universities with graduate programs) so make sure you do your homework about what's possible or stay with safe suggestions (e.g., everyone would probably be interested in someone helping out with Psi Chi).

- starting a Master's program in your subarea of the field
- creating a new major or concentration in your subarea of the field
- incorporating statistics into a lab course or bringing statistics into your department to create a research and statistics sequence of courses
- applying for a summer research grant with other faculty
- if the faculty alluded to taking something off their hands (e.g., Psi Chi, website, social media) in the ad or in the phone interview, have an idea about it
- mention implementing group or peer advising to help with the burden of advising in your department
- have an idea about incorporating research into your teaching load by offering upper level courses in your lab

Be likable. In small departments, they are often basically hiring a new friend. Be easy going but serious, respectful but not uptight, flexible but ready with answers, and be very appreciative of their time.

Related to the above point, fit is a huge deal at teaching institutions. The search committee is often trying to determine if you will be a cooperative, rather than competitive colleague. The search committee will not be interested in folks with stellar research records if they come across as having the potential to be a very difficult or unpleasant colleague.

Ask to meet with students if you aren't already scheduled to during the campus interview. At a small liberal arts school, meeting with students is just as serious as meeting with faculty. Ask the students what they are looking for in a professor. Ask them about their relationships with professors. Remember, they will tell the faculty everything that you said so stay professional at all times.

Know your weaknesses and have a positive spin on them. The search committee is likely to ask you about your weaknesses or any concerns you have with the position. Be ready to be honest and say something (nobody is perfect) but also have a positive spin on it. For example, that you can be so focused on students that you will have to remind yourself to take time for research but you will be vigilant about doing so because you know that your research program ultimately serves them too.

Some other questions you can ask in down time are as follows—there is always down time when they want to know if you have any questions and if you don't have questions, again, you appear uninterested.

- What are new faculty members surprised by when they arrive?
- What do you think a new faculty member will struggle with?
- What particular assets are you looking for from the new hire?
- How does this department compare to others across campus?
- What schools are your competitors?
- What schools do you seek to be like?
- What are the department's goals for the next 5 years?
- Is there a sabbatical?
- What is the teaching load?
- What are your department's grad school acceptance rates?
- What is the departmental budget?
- Are there travel funds for faculty?
- Is there grant application support, such as a Research and Sponsored Programs office?

CLASSROOM RANGE

There is one interview question that is commonly asked and doesn't have a great answer. Contemplating your answer to this question will be helpful in preparing both your teaching statement as well as for an interview. The question is usually some variant of, *We have a student body with a wide range of ability, from students who are too advanced for the class to students who should not be in college. How are you going to deal with that in the classroom?* The reality is that every classroom, indeed every group of students, will be have students possessing a variety of abilities. While this spread may be greater at some institutions than others, it can also vary from class to class, sometimes dependent on the curriculum. For an example at the individual course level, I have taught an Introduction to Cognitive Neuroscience class that was one of three introductory neuroscience courses in a major. Therefore some students had already completed two full courses in neuroscience and some had never taken anything related to neuroscience. That's a pretty wide spread of ability simply as a function of curriculum. At the institutional level, some students in your class may be attending the school on full academic scholarships, meaning they were wildly competitive, while other students barely got into the school.

Reaching the most students across a wide range of ability in the classroom is something I don't claim to have mastered. However, some thoughts on how to address this include assessing ability on day one to know exactly what kind of distribution of ability you're dealing with, learning your students' names so you can reach out to students who are both struggling and exceling in class, emailing students after exams who are at either extreme of the grade distribution and encouraging them to come and chat with you about either how to better study or how to be more challenged, and recognizing that there is always going to be this spread and that your goal should be to reach the majority of students while hopefully also spending some time outside of class (ideally during office hours) challenging the students who are advanced and supporting the students who are struggling. Perhaps have some ideas about how to make attending office hours more rewarding. I know of professors who have office hours at different places on campus, such as in front of the fireplace at the student union, in order to try and encourage more students to attend office hours. If you get this question, or some variant of this question during your interview, I encourage you to answer with some thoughts on the topic like those I have mentioned above and then ask how the faculty on the search committee handle the spread at their institution. Remember that not having questions for your search committee makes you appear uninterested and bouncing a question or two back at them when you don't have a fantastic answer is a way to show you respect and acknowledge their experience and are willing to learn.

NEGOTIATION

Most of the advice applicants get regarding negotiation comes from their faculty advisors, who are at research institutions and don't know what to expect in terms of salary and startup at teaching colleges. Generally speaking, there is less variability in the negotiation process at a liberal arts school than there is at a research-intensive school. However, this can certainly vary based on a particular school's resources and budget. At a liberal arts school, you can expect startup funds to range from a few thousand dollars to tens of thousands of dollars, and this will vary based on the specific institution and the type of research you do. Of course, at a research-intensive school, you're talking about startup packages in the hundreds of thousands of dollars, and around a half million dollars is relatively standard. These are just rough estimates, but the variability of this range is why I think it's wise to reach out to the search committee chair and inquire about funding before submitting application materials. Doing so also gives you an idea about what is possible during negotiation (but see the interview at the end of this chapter for a different perspective). Of course, I think it's wise to ask for more than the initial offer when negotiating your salary and startup package after you have an offer on the table, but don't be surprised when there isn't much movement in negotiations for these teaching positions. However, it is always true that the time to ask is before signing your acceptance letter.

IS A NONTENURE-TRACK JOB SO BAD?

Let's pause this discussion of landing a tenure-track job at a liberal arts institution and contemplate non-tenure-track jobs. Although the specific labels vary across institutions, tenure-track jobs are those that come with job security following a certain period during which faculty members prove themselves worthy of job security. Non-tenure-track jobs are those that do not formally come with long-term job security. There are certainly other distinctions between the two that may result in non-tenure-track employees being second-class citizens, such as often not having the right to vote in faculty meetings or senate. As academics, we have had it beat into our heads that tenure is the golden ring dangling above us and that once we have tenure we will live happily ever after. As two faculty members who have gotten tenure, we want to tell you that you are going to be sadly disappointed if you think that tenure is going to make you happy or make your life magically easier or make your days qualitatively better. One of the first things that happens after you are granted tenure is you get more service work plopped on your desk! So let's entertain some of the main criticisms of non-tenure-track jobs and then consider some of the upsides.

Many folks will argue that non-tenure-track jobs are awful because you might not have enough courses to teach, you don't have input into your teaching schedule (either the times courses are offered or the actual courses you are teaching), your salary is not enough to live on, your contract is year-to-year, et cetera. I am certainly not trying to downplay or undervalue those concerns. A very real issue, a crisis even, facing academia is how some institutions are taking gross advantage of non-tenure-track employees and not paying them livable wages.

While I do not want to defend deplorable working conditions for some adjuncts, I do want to argue that, at the right institution and under the right circumstances, a non-tenure-track job can be a wonderful career choice. As someone who has always worked at institutions with high teaching loads (12 credit hours per semester), I recently left my position as an **associate professor** for a position as a lecturer at Ohio State University, where I spent a year before I was promoted to a visiting associate professor position. In the US, a lecturer is a non-tenure-track position. Here are some bonuses of this move that I experienced. First, the gift of time—where did all this time come from?! When you are not tenured or on a tenure-track, you are, in many ways, an invisible employee.

I know this is problematic at some institutions, but at others it means the gift of time. As a non-tenure-track professor you may have essentially no service work; that means no committees, no advising, no department meetings. This frees up hours and hours and hours of time to conduct research, prepare lectures, and write books (if you are holding a book in your hands right now, I am proving my point)! You may still get funding for travel (as a lecturer I received the exact same travel stipend from the department as I did in my two previous associate professor positions), be able to conduct research by taking advantage of the same subject pool tenure-track professors use, have laboratory space and research resources just as good if not better than some tenure-track professors get, and finally write up some of those manuscripts that were collecting dust on your hard drive while you advised hundreds of students, went to department meetings, served on search committees, met with the curriculum committee, et cetera. I want to be really clear here that not all non-tenure-track jobs are this luxurious and although I was fortunate, not everyone is so lucky. However, I can't in good conscience write a book about advice for the budding scientist and not break the mold to argue against the idea that tenure = happy or tenure-track jobs = the gold standard of academic jobs. There are many paths to happiness and I don't want everyone to balk at non-tenure-track jobs when they might be ideal for some applicants at particular points in their lives.

CONCLUSION

Faculty positions at liberal arts institutions are competitive. Many of these schools are continually updating their goals and expectations to competitively meet the needs of students (see the interview with Glenn Sharfman in Chapter 8 for what administrators value in their faculty at teaching schools). While the dynamic requirements for teaching positions can be frustrating for the applicant, once in a faculty position, the benefits of working closely with appreciative students makes for a very rewarding career. As students, Geoff and I both attended Big Ten schools (large, public schools) for undergraduate and graduate school. As faculty members, we have collectively worked at a wide range of institutions from a small liberal arts school to a Big Ten school to a mid-sized state school to a very prestigious private university. We have been very thankful for our opportunities at all of these institutions but are likely to encourage our children to seriously consider attending liberal arts institutions because of their individualized focus on every single student. The struggle you may feel as you navigate applying for faculty positions at these schools will pay off infinitely when you work closely with students who you will ultimately cheer on for a lifetime (say hello to wedding invitations, baby showers, pictures of growing families on Facebook, and more). Your students from liberal arts institutions are in your life forever. It's a wonderful life.

> **Ashleigh's Interview with Lauren Hecht, PhD**

homepages.gac.edu/~lhecht

Lauren Hecht, PhD is an Associate Professor of Psychological Science at Gustavus Adolphus College, Saint Peter, Minnesota. I know Lauren from the Cognition and Perception PhD program at the University of Iowa. Lauren is the perfect person to interview for how to get tenure at a liberal arts school because she loves teaching, interacting with students, and mentoring junior faculty. She has taught nine different classes at three different institutions, mentored dozens of students in her lab, served on faculty search committees, published in the field's top journals, and much more. Lauren has earned tenure at a prestigious liberal arts school so I think her feedback on how to get a job will broadly apply to, if not over prepare you for, most teaching institutions that have job openings.

SUGGESTIONS ON GETTING TEACHING EXPERIENCE

Lauren added to some of my ideas on getting teaching experience based on her own experience. First, she suggests trying to teach summer or evening courses while you are in graduate school. These unique opportunities might allow you to teach something other than introductory courses often offered for graduate students to teach during the school year. It might also be worth seeking teaching opportunities outside of your department. For example, you might be able to teach a statistics course in the math department or a research methods course in the communications department. Lauren remembers one applicant who taught an afterschool workshop. She explained that you can spin lots of experiences into teaching experience. Lauren described how much search committees love it when people reflect on teaching experiences, even if they failed, provided they tried something new and were seeking ways to improve. It's not enough to say, "I thought about this experience and here's why it went well/poorly." Talk about what you did or will do next time around, especially when something didn't go as planned.

COVER LETTER

Lauren couldn't agree more about the importance of the cover letter. She reiterated my point above that they are the best place to get information and they drastically vary in terms of their effectiveness. Lauren said that the cover letter needs to be accessible because the expertise of the faculty varies so widely, even within a department. One way to capitalize on a cover letter geared toward nonexpert colleagues is to highlight what to look for in your application materials. For example, briefly describe your research and direct them to see your CV and research statement by saying, "I've got several publications on visual attention in well-respected journals in the field, including *Journal of Experimental Psychology: Learning, Memory, & Cognition* and *Attention, Perception, & Psychophysics*." Don't neglect to mention your research in the cover letter.

Despite whole-heartedly agreeing with the importance of the cover letter, Lauren did emphasize that you don't want a three- or four-page cover letter. You need to highlight your many qualifications and convey that you're enthusiastic about the position, but you could easily end up with a four-page cover letter that nobody wants to read. Strike a balance between information and length by using the cover letter as a space to highlight what the search committee will see when they read the rest of your application materials, making sure the summary is clear to a nonexpert. For example, say, "I've mentored ten undergraduate students and I took one to a conference," or "I've done a lot of work with undergraduates (see CV and research statement)."

In order to rise to the top of the applicant pool, make sure your cover letter expresses an interest in the geographic location of the school (which might be more important in potentially unappealing rural locations), provides specific examples to emphasize your point, is clear for nonexpert members of the search committee, and shows that you understand the community you're coming into (e.g., mention your draw to the liberal arts). You have to be able to say why you want to be at a liberal arts school. You need to say more than you just want to teach; you need to understand what liberal arts institutions have to offer and make sure you can communicate your desire and understanding to the committee. Lauren also points out that some of these suggestions (e.g., clarity, providing examples) are advice that applies to all the materials, not just cover letters, and also apply to interviews (e.g., specific examples, why you want to be at a SLAC). For instance, some specific examples are better placed in the teaching statement.

QUESTIONS YOU WILL BE ASKED IN YOUR INTERVIEW

Lauren described how her department had an agreed upon set of questions they asked candidates during phone interviews. This promoted fairness and equality across interviews. Most schools that are interviewing for teaching faculty do not have money to bring lots of candidates to campus, so you really want to nail the phone interview if you get one. She shared the gist of a few of the more compelling questions committees ask and also described solid answers. As a general overview, you can expect questions about two of the three pillars of tenure if you are interviewing for a tenure-track job (i.e., teaching and research, but probably not service). Rather than asking about service, Lauren says that in their interviews they focus more on the mission of the school and fitting into a SLAC community, which she points out are not mutually exclusive topics, of course. When an institution has prespecified their questions for equality and legal issues, they will not be able to ask specific questions or follow-up questions, so be ready to give strong answers to the following questions.

A teaching question might be *provide an example of a time when you found a clever way of motivating students in one of your courses* or *give an example of an adversity you've faced interacting with students in the classroom and how you overcame it*. These questions are best answered by demonstrating that you are a reflective teacher, someone who is going to put a lot of effort into your teaching. They are looking for an understanding of how student and faculty will interact with one another, rather than issues like how hard it is to set up a course website. Lauren agreed this is a version of the question about how you are going to reach everyone in a classroom that has a wide range of ability. If you can highlight this in your teaching statement, that's even better because it might get you closer to the short list of folks who end up with interviews. However, Lauren says don't feel like you have to save it for the interview. In your answer you can reiterate and elaborate upon information that your application materials contain (which can oftentimes be viewed quite favorably by a search committee). What the committee is looking for is that you have considered the possibility that you might have to deal with students who are starting at different places. A bad answer uses the push-ahead mentality of *I'd make clearer PowerPoint presentations* or *I'll make sure I have regular office hours*. They really want to hear that you will change your pedagogical approach to meet the needs of the class. For example, propose pairing up students to take advantage of peer-teaching and discuss evidence that suggests peer-to-peer learning can serve both the more advanced student because they are teaching and the less advanced student because they are learning from a peer who might better communicate with them. A good answer here means going beyond what is already planned for the course, demonstrating reflection, interest in teaching effectiveness, willingness to try something new, and flexibility.

Another version of this question is: *What happens if you face adversity in a classroom?* Bad answers to this question would be describing what you should already be doing, like holding office hours. What you should demonstrate is that you've thought about the fact that you need to go beyond office hours for struggling students. For example, you might explicitly bring a student to your office to talk and help put them in touch with specific campus resources related to their problem. This might be better if the adversity you faced was related to only one individual in the class. This point gets to the fact that you should help clarify the adversity you are addressing before saying how you will handle it. The adversity must be clearly stated because the committee would want to see an appropriate response. For example, calling someone to office hours or doing a one-on-one intervention might not be sufficient if you have a pervasive issue (or one that might affect many students in the class), such as lack of preparation in reading journal articles, thus not being able to successfully complete a more complex exercise that relies on this skill, or an issue like widespread discomfort from an altercation in class. For example, one of my students was offended by a peer during what was supposed to be a friendly debate, made a clear statement of this fact, and walked out of class while everyone watched the entire exchange. However, in instances when the adversity is student-specific (e.g., a student with a learning disability, a student struggling with content due to personal experiences, discord in a peer-group), then a personalized response is entirely appropriate. This would be a great time to acknowledge that you will use advising as a means to supplement teaching. Indeed, you can leverage advising in certain ways to help enhance your application because all faculty members will have to advise and it might be good to demonstrate that you know that. You could follow up by asking when you're expected to start advising and whether you will advise undeclared majors in addition to majors in your department. Lauren adds that this is great for a campus interview, but she wouldn't waste too much time on it in a phone interview, unless you're really worried about it.

    A question about your research might be, (1) *How do you see your research fitting into our department and what items would you need to do your research here?* or (2) *What research would you like to do here and what would you need to do it?* An ideal answer to this question would include a description of how you would fit undergraduate students into your research program. You will of course talk about what research you've done and where your research program is going, but the committee is also interested in hearing about how you are going to involve the students in your research, your needs for space, and what is and is not feasible. This question also helps the search committee determine whether or not the candidate is picturing their research program in the context of the department. Has the candidate done their homework on what research is already happening? If so, then they would have a sense of what may or may not be feasible. What gaps could the candidate fill (e.g., you'd be the only person conducting psychophysics)?

    Lauren also made the point that search committees want to know how you'd deal with issues related to sampling (i.e., acquiring a specific population) and measurement. She once saw a candidate apply to a rural SLAC who wanted to work with schizophrenic patients. The local population could not meet the demands of the research, so how would the candidate handle that? Likewise with a scanner,

Lauren said they once had someone apply that needed costly equipment like an MRI scanner. Although the solution might seem obvious (collaborate with someone at a large university), the committee was concerned with how often they, students especially, would see this individual on campus if they were always traveling to another school. And if data are always being sent from another institution, what research experiences are the students receiving from this individual?

Another question might be, *If you were sitting down with a prospective student and their parent, how would you describe the benefits of a liberal arts education as opposed to other, cheaper options?* The best answer to this question would be taking this opportunity to specify that you understand the worth of the liberal arts education. Lauren said that if you can answer this question well, you can answer a lot of related questions well.

Another question that Lauren said committees often asked of faculty candidates is, *What classes would you like to offer that are new to the curriculum?* Be prepared to answer this question. They are specifically looking for classes that are not mentioned in the ad and not currently offered at their institution. Ads often list courses that must be taught (e.g., introduction to psychology, research methods), so the best answers need to go beyond the ad.

This is a chance for you to demonstrate that you have done your homework on their curriculum. It might be even better if you can integrate your proposal into some unique aspect of the academic calendar like Janterm or Maymester, possibly with experiential courses or service learning, a first year seminar, or into a general education requirement. You could also demonstrate that you have researched the *community* (a buzz word at liberal arts institutions) by suggesting co-instructing a course with a professor in another department.

QUESTIONS YOU SHOULD ASK DURING YOUR INTERVIEW

It's important to ask about the dynamic of the department. You can do this in multiple ways. For example you can ask, *How does everyone in the department get along?* or *What are department meetings like?* Any question that gets at the camaraderie of the department is a question you can and should ask of multiple faculty. Lauren says that some suggestions, such as this one about the climate of the department, you can ask over and over again to see if you get the same answers when meeting one-on-one with different faculty members. Lauren and I have both interviewed at institutions where some people gave one answer and others gave a very different (and telling) answer.

Some other great questions that demonstrate your interest in the position include asking about the town and the surrounding community, inquiring about the campus community and atmosphere, asking what there is to do in the community for fun, and inquiring about what equipment you will have access to. These questions demonstrate that you are thinking about what it would be like to live there and can be a nice break from other types of interview questions. Lauren remembers asking whether faculty get together outside of work to get a pulse on the departmental relationships and demonstrate that she was seriously considering how she would fit into the community. Lauren remembers asking this question broadly, of many different faculty members, and heard mixed results. Sometimes she heard what department colleagues were doing, but it was also nice that people sometimes mentioned what faculty from across departments would do (e.g., we have a communication studies prof and an English prof who would often coordinate semiregular wallyball games that involved interested faculty from a variety of disciplines).

Lauren also suggested asking if and when sabbaticals are offered, following up on my suggestion of asking whether there is a sabbatical, because some schools have pre-tenure sabbaticals, whereas others offer only post-tenure sabbaticals. Another telling question would be, *Are there separate departmental budgets for teaching versus research?* The answer to this question tells you what you'd have access to for teaching versus research. They may not want to give you specific numbers, but you will get an idea of the continual support offered by the department and it demonstrates that you would like to accentuate your teaching if possible with funds for an activity like a cow eye dissection.

WHAT YOU CAN BE DOING DURING GRADUATE SCHOOL

Along the lines of not having your career perfectly planned out, Lauren told me that as a first generation college student, she had never considered teaching at the college level to be a possible career. When she thought of teaching, Lauren didn't think of teaching at the college level. Rather she thought that teaching was for grades K–12. Lauren went to graduate school because she was doing a thesis as an undergraduate student and she enjoyed research. She actually had a rough transition to graduate school, going from a small liberal arts college to a Big Ten university, and was thinking about quitting the PhD program at the end of her first year. Serendipitously, Lauren had to co-teach a course during the summer after her rough first year and she fell in love with teaching. This opportunity put her on a path to seek interactions with undergrads in mutually beneficial relationships, such as mentoring students in the lab who were collecting her data. In seeking out the things enjoyed, she was actually positioning herself to secure a highly coveted tenure-track faculty position at a premiere liberal arts college.

When I asked Lauren what someone could be doing during graduate school to better position themselves for a teaching job, she clarified that the best thing is getting actual teaching experience. If you can't teach, as I mentioned above, some graduate programs offer a graduate course in teaching or if they have a center for teaching they may offer a certificate in teaching. Lauren also mentioned the opportunity to give lectures or workshops for groups like the Midstates Consortium for Math and Science[1] (which also includes psychology in their educational repertoire). At the regular meeting of the Midstates Consortium for Math and Science for Biological Sciences and Psychology, graduate students give demonstrations or educational presentations, usually related to neuroscience. Lauren says that even details about those experiences could enhance your application. Although hands-on demonstrations like dissections with a Girl Scout troop are not classes, they are still events for which you planned the materials, you can describe what you taught, and you reflect on what went well and what you would change in the future. All of these options would give you meaningful experience to talk about in an application.

Other options Lauren discussed are taking a teaching position when you are ABD, which stands for *all but dissertation* (i.e., you have completed all the requirements for your PhD except your dissertation), or a true teaching postdoc. Lauren's institution offers teaching postdocs, formally called postdoctoral fellowships, where individuals teach one course a semester and are expected to spend a significant amount of time on research and advising. The primary goal of this position is to augment teaching experience for underrepresented groups. Lauren added the caveat that this position has been primarily available in the humanities at her institution. Departments apply for university funds to host the fellows, and her department just recently started talking about which year they might apply. Offered at many institutions, these positions offer a chance to gain teaching experience, garner letters of recommendation from a variety of faculty, and allow time to write up your collected data, which enhances your research portfolio.

CAMPUS INTERVIEWS

Lauren wanted to make sure that readers understand some issues unique to on-campus interviews. In particular, some interviews will involve teaching demonstrations where you teach a class, some involve research talks, and some involve a teaching research talk, where the school expects you to teach undergrads via your research talk. These demonstrations are make-or-break moments. Your talks are a very important chance to demonstrate that you can

---

[1] https://mathsciconsortium.org

speak to a variety of audiences. If you do not know who the audience is going to be, you need to ask in advance. When the audience is current faculty members, you need to keep in mind that it really just means you are speaking to college educated people, or basically fourth year undergraduates, because they do not have expertise in your topic area. Be certain that you pitch your talk, even your research talk, to the right level.

If you're asked to do a teaching demonstration, that's when you could try to demonstrate the pedagogical methods you discuss in your teaching statement. If your application talks about active learning, try to demonstrate active learning in your teaching demonstration. Whether you give a research talk as a simple research talk or as a teaching research talk, be sure to step people through every aspect of your talk. For example, don't put up a graph without considering that the audience doesn't know what they are looking at; rather, orient people to the x-axis and y-axis on your graphs. Just like when they are reading your research statement, the search committee is thinking about how your research fits into the mission of a teaching school. If you're not accessible in your research talk and statement and your colleagues can't understand what you're doing, then they will determine that you will not be able to work well with the undergraduate students.

KISS OF DEATH

Based on her experience on search committees, Lauren came up with a short list of things you should definitely be cautious of when you are trying to get a job at a teaching school.

- Don't be dishonest in your application materials. For example, do not masquerade non-research publications (e.g., blogs) as actual peer-reviewed research publications. Blogs are not publications but some applicants try passing them off as research. Another example would be talking about a teaching method you would never use because it doesn't work for you. For example, don't talk about community-based learning if it's something you would never consider doing. Clarity of materials is essential.
- Don't neglect to talk about involving undergraduates in research. If you don't talk about undergraduate involvement in research, you're going to fall lower in the pile of applications. When applicants are strong in research but have little or no teaching experience and the application materials leave it entirely ambiguous how research, which is their strength, would fit into working with undergrads, it does not help their cause. It is crucial you acknowledge how your research program serves undergraduate students.

- Don't overemphasize your connections with researchers at other institutions. A major concern with candidates who parade out their collaborations with faculty at other institutions as a major strength is that they will never be on campus. Unless the collaborators are within walking distance, collaboration takes your time and resources away from campus and prevents students from participating in your research program. If data is being collected elsewhere and being sent back to you, that is ok and should be made clear in your application materials. However, you should consider carefully what the research experience will look like for undergrads. Will they be able to assist in data analysis or is it too complex to learn in the relatively short period of time students have available to work in labs during undergrad? What research skills will students learn if the research is not carried out on campus?
- Don't be vague in your teaching statement. Lauren says at some institutions you could even get an interview if you've never taught before, provided you put substantial thought into your teaching statement and provide thorough examples. Anybody can say *I'd use a flipped classroom*, so you need to give an example of what the flipped classroom would look like, such as *Here's one example of something we could do in class to supplement an online lecture. I'd have students get into groups and ask them these questions about A, B, and C, and they'd be able to demonstrate X, Y, and Z*. Rather than make vague statements that anyone would be able make, show that you've thought carefully about teaching even though you don't have experience.
- Don't fail to explain that you understand the liberal arts. Search committee members wonder whether applicants really understand the liberal arts environment and what it means to be there. Communicate that you understand the school's mission and what the faculty do. Explain why you want to be there and how you fit in.
- Don't fail to ask questions. Lauren reiterated the sentiment that you should have a list of questions ready. She remembers interviewing one person who never asked a question during any meeting. Not asking questions makes it sound like you're not interested.
- Don't appear unable to get tenure by any of the metrics described above about what these schools are looking for in their successful faculty. A tenure-track hire is an investment and the tenure piece is important because the faculty doesn't want to invest time and money into someone who will fail. At small liberal arts schools it is possible to have a tenure line taken away from an individual department, so if someone fails to get tenure and leaves, that could hurt the entire department. Since the department wants whomever they hire to get tenure, show your interest in the process and your ability to achieve.

NEGOTIATION

Lauren thought I was being too ambitious in telling you to contact the department chair or head of the search committee about the amount of startup funds, because in her experience it's usually only the Dean or Provost who has that. I also suggested that you only propose needing the research equipment they tell you is financially possible, but Lauren disagrees saying that it is too cautious to only propose research needs that meet a budget (when you finally get that number in hand). Lauren said she asked for more money than her chair thought was possible for equipment and startup funds for her lab, and she got it.

Lauren's got a really great point here about negotiation. The degree to which there is room for negotiation might really depend on the school's financial situation. Although the interview is a good time to gauge the financial stability of the school (and asking about it is a great question), it is wise not to sell yourself short in pledging to be inexpensive. My advice stems from my experience seeing that applicants won't get a second look if they report needing research equipment the college could never afford. You might want to have several versions of your research statement, with each proposing different needs if there are significant differences in the finances of the institutions to which you are applying. You need to make it really clear if there are things you just can't live without (e.g., if you have to have an electrically shielded booth, for example), so be upfront about those costs. Of course, also be wise about not selling yourself short and asking for too little.

CHAPTER 7

How to Get a Job at a Major Research Institution
by Geoff

For many of us, the goal of graduate school and our postdoc was to get a job at a major research university. This is because our mentors had such jobs and those jobs looked cool. Large research grants allowed our mentors to travel to conferences and do exciting science. The grants meant the lab could keep churning out exciting papers and chasing new ideas. Also, many of the awards from societies and organizations are for research, with teaching awards being internal to universities and more rare. Given all of this, faculty positions appear to be the path to fame and glory in academia.

First, I want to elaborate on a bit of previously introduced terminology. An R1 university is one where research is priority one. The term R1 comes from a Carnegie Foundation system developed in 1970 to classify institutions of higher learning to compare similar universities. R1 schools are those that grant doctoral degrees and have the highest levels of research activity across universities. Many of the major universities with football teams that play in bowl games are R1 universities (Stanford, Michigan, Vanderbilt, Northwestern, Ohio State, University of California Los Angeles [UCLA], the University of Texas, Austin, et cetera), plus the Massachusetts Institute of Technology (MIT), California Institute of Technology (Caltech), some of the Ivy League schools, and a number of others. In faculty jobs at R1 schools, success is defined by your research productivity.

Let me just take a moment to tell you why R1 jobs are so sought after. The first reason that these are sweet jobs is that teaching loads are generally pretty light. This might sound heartless after Ashleigh's chapter on teaching but some of us love research more than teaching. At many R1 schools, you will teach zero to three classes per semester, depending on how much grant funding you have for your research. How do you teach zero classes? You can buy out of teaching using your research grants at many of these schools in the USA. You pay the college to hire someone else to teach in your place. Some schools do not allow this, but then they typically have consistently low teaching loads, such as requiring one class per semester, assuming you maintain an active program of research with external funding.

The second reason jobs at R1 universities are sought after is that they come with significant startup funds. When you negotiate to take a job at an R1 institution, much of what you are negotiating is how much startup money you will get. This is enough money to buy all the equipment you need and run the laboratory for 3-5 years before you get external funding from a federal agency, private foundation, or corporation (i.e., a grant or other research funding). As you can probably guess, startup funds can be fairly large amounts of money that allow you to outfit your lab with the newest and most powerful equipment available. These funds also allow you to hire research assistants so that you can have staff immediately.

The third reason these are highly sought jobs is the freedom they offer. When you have a faculty position at an R1 institution you are allowed to study anything you want in your laboratory, within ethical guidelines. When is that possible in the real world? If I work for General Motors they are going to require that I make cars. If I work for Pfizer I need to make or test drugs. However, in my laboratory I can study anything I want. The Howard Hughes Medical Institute provides grants to enhance just this kind of flexibility, allowing a physicist to become a cancer researcher, or a neuroscientist to become a computer scientist. These are hard grants to get, but they emphasize an implicit aspect of R1 faculty jobs: they are the ultimate in intellectual freedom.

Is that enough telling you why these jobs kick butt? You probably already want one because this is what your academic mother or father did. But I wanted to make sure your appetite was sufficiently stimulated.

HOW TO GET AN INTERVIEW

Having convinced you that getting an R1 job is worth the effort, you naturally want to know how to get one. This is where I hit you with the brutal reality that these are hard jobs to get because they are so sweet.

Search committees for faculty jobs routinely receive 100–500 applications for a single faculty opening. These applications are all from PhDs. Sometimes applicants already have a faculty position and are looking to move up. From this huge pile, search committees with come up with a list of 3–5 people to interview in person. Then they will pick one to hire. Odds of 0.02–1.0% don't seem great. However, you can totally do it! Getting one of these jobs requires dedication, tenacity, careful time management, and a constant eye on your CV, but they are completely obtainable.

Your application for an R1 job rises to the top of the stack and you get an interview for one of these jobs by having one of the best-looking CVs. *Best looking* is predominately defined by number and quality of publications, as we have described in previous chapters. The department wants to hire an independent research scientist. To do this, they need to know that you know how to write a paper. You demonstrate this in your CV with your many, high-quality papers. They need to know that you know how to obtain grant funding. You demonstrate this by getting a grant as a grad student or postdoc to fund your research. They need to know that you can teach and mentor students. You demonstrate this by your job talk, your teaching experience and mentoring of undergraduates and junior graduate students in the labs you've worked in.

You can figure out what is important if you play the game of imagining what it would be like to be a member on the search committee. You already know that you are judged by the number and quality of your publications, your grant funding, and the opinion of your work by those in the field (e.g., citations, letters of recommendation). These same metrics are used when considering your application.

So how can you get an interview? Publish important papers. Apply for grants as a grad student and postdoc. Mentor people junior to you. Guest lecture when opportunities present themselves.

You may wonder which of the above suggestions is most important for making the short list for an R1 faculty job. The truth is that different people on the search committee have different biases. Some focus strictly on previous publications, some think you need a grant on your record, others really focus on the research statement and what you say you are going to do in the future. Let's talk about the ads first, and then the application materials.

THE ADS

I would like to caution you against trying to overthink job advertisements. If you see an ad description that you seem qualified for, then you should send in your application if it is somewhere you could see yourself going. Don't make the mistake of going to the department's webpage, seeing people who do something related to your work, and thinking that the school won't be interested in you because they already have people like you. You cannot know this a priori. Different departments will have different philosophies with regard to hiring. Some will want to build on strength. If this is the case, then your work having some overlap with existing faculty members will be an asset. In contrast, other departments may be looking to hire for breadth of coverage. It doesn't cost you anything to apply. So send in your application for the job if you'd like to be there.

On the flip side, don't take it personally if you don't get invited to interview. They may be looking to replace someone who is retiring and studies a specific topic, but wanted to be open to opportunity, so they wrote a general sounding advertisement. You also have no idea what the rest of the applicant pool looks like. Just apply to all the jobs that you could imagine yourself taking and don't overthink what a particular department or search committee might be looking for. Having choices requires you to apply broadly, so do that when you feel you are ready.

PARTS OF THE APPLICATION

The part of your application materials that you've been working on for years is your CV. When you are getting ready to send your CV in as part of an application, spruce it up a bit. Here are some specific tips.

Put your education at the top, then honors and awards if you have them. Next, list your publications. If you have papers in preparation that you want to list, only list those for which there is a manuscript that you could send. If you have 18 published papers and have 15 listed as in preparation, this does not help you. While including submitted papers is important, listing projects that are in preparation looks like you are padding your list, or you are bad at closing the deal and getting mature projects written up and submitted. One of my mentors told me to delete all in-preparation papers from my CV because they thought such papers are meaningless. I kept the two or three that really had drafted manuscripts.

After publications, list any grants you've gotten. This should include the total of the actual funds received. This allows search committee members to understand whether it was a small summer grant or something that paid your salary for several years. If you helped your mentor prepare a large grant (e.g., an R01 to NIH) it is okay to list this experience, provided you can concretely state what you did for the application and the breadth of your knowledge of the parts of the application. If you submitted a grant that was not funded, you can include that as well since any experience is better than no experience.

Next, list your teaching experience. Highlight any courses you've taught as instructor of record. Being an instructor of record means that you receive student evaluations of your course. If these are very positive, then it can be useful to include them in your application materials so that the search committee can see how the students liked your class. Although teaching experience is most important for the teaching jobs covered in Chapter 6, you might be surprised how much teaching matters at some R1 institutions (see interview with Dr. McNamara at the end of Chapter 9).

After publications, grants, and teaching experience, you can list conference abstracts, people you've mentored, and the rest. The general idea is to put the important stuff at the top.

You will also send reprints (pdfs these days) of your three to five best papers, depending on how many the ad requests. Think about which of your papers are in the highest impact journals. If you have work demonstrating your independence, such as a sole or senior author publication, those are excellent to include. Typically sending commentaries is bad form as these are not empirical papers or theoretical ideas that came entirely from you. Instead, they are a reaction to other peoples' work. If you have a paper that is going to be the springboard for what you are going to do in your new lab, that is good to include. Explain in your research statement why you selected those particular papers. Your current mentor can help give you guidance on which papers to send. In general, you want to send your strongest publications in the best journals that show the committee what you do and where you are going.

Now for stuff you have to write. You will need a research statement. As I just mentioned, this typically builds on some of your recent work. The search committee will want to know that you have the skills you need to do the research you want to do. If you are proposing to do something completely new that you have not been trained to do, it will cause concern. The best research statements are like good sentences: they start with something the reader already knows and then tells the reader something new. That is, tell them about the line of work that you have already established, then tell them what you propose to do next. For some people, the research statement will have multiple sections, with each highlighting a different line of work and where they propose to take it next.

The research statement should not just rehash what the search committee members can get from your papers or your CV. That is, it shouldn't just review what you've done. The search committee members who emphasize the importance of the research statement like it because it tells them where you are going. If you have begun writing your first grant, use the research statement to tell the committee this and to highlight what you propose in the grant. You can include preliminary data figures from that grant proposal in the research statement. The research statement should layout your vision for your work during the next 3–5 years.

The other statement that you need to write is your teaching statement. For many applicants to R1 jobs, this is the harder of the two statements (i.e., teaching and research) to write. It helps if you can say that when you taught your own class you did A or B and it worked well, but it is really not essential. If you have not taught, you do not need to say that. What a search committee wants to know is your plan. Do you plan on running a flipped classroom? Do you plan to teach with journal articles instead of textbooks? Do you want to teach students to write, evaluate research, think of new experiments? You get the point (see Chapter 6 for more advice on teaching statements).

The teaching statement is a good place to declare what you would like to teach. If you would be willing and able to teach statistics, experimental design, or introduction to psychology, then highlight this. These are classes that almost every department of psychology or neuroscience struggles to staff. So saying explicitly that you are willing to teach one of them shows the department that you are willing to be a team player and cover a course that is required for undergrads across campus, not just in the psychology department.

Almost all universities have their course offerings and requirements for majors available online. It would be wise to look at the list of required classes and see how you could contribute by teaching one of these service courses, that is, a required class that the department must offer. I fell in love with teaching Intro to Cognitive Psychology. It wasn't my first choice, but I found that it improved my research perspective and helped me recruit excellent undergraduates to work in my lab. At the end of a teaching statement people often describe their mentoring philosophy, since training graduate students is a part of teaching. This is a good place to make it clear that you've thought about this aspect of the job.

Spend your time really tuning up these statements. Ask your current and previous mentors to read them and give you comments. Proofread them carefully. A beautifully written statement shows that the candidate can write well generally.

THE JOB TALK

When you submit job applications, you should get your talk ready right away if you have not already done so. Some schools will move quickly, inviting applicants to talk in a few weeks, others will invite people months later. Ideally you will give multiple practice talks with your mentor and current department. The goal is to have a very strong job talk.

You need to be able to give an engaging and clear talk because your teaching ability is judged, in part, by the clarity and coherence of your job talk. For this reason, I tell my postdocs and senior graduate students to begin thinking about how to package their work into a job talk well before they hit the job market. You cannot start on this too early.

In your job talk, you want to tell the audience about your central line of work. If you have a couple of lines of work, a successful strategy is to think of a general question or idea that links them. Then tell the audience the direction your research will go in your new lab, should you be hired .

A common error is to try to cover too much. That is, people try to cram 10 papers into a 50-minute talk. The goal is not to tell the listener about everything you have ever done. Instead, the goal is to tell a story.

Remember that you are telling your story to an entire department. Psychology and neuroscience are broad fields with people who study everything from low-level pitch perception to mood disorders using everything from patch clamp measures of ion channel activity to magnetic spin sequences to measure activity in large chunks of tissue. You want to engage everyone.

Have a good hook that can help everyone understand why your work is interesting and important. Start with some real-world situation that leads to a general question. I study visual search, so I like to start with some pictures of a bunch of objects in which you are trying to find one thing. Then I say that what we try to understand in my lab is how do you find that one thing in these cluttered scenes. A grade school kid could get it. That's good. Then you can get a little bit more technical. Provide an initial answer to that general question with a specific experiment or two. Then tell them where your logic took you next. Link the work with some logical framework that makes your program of research look systematic, cohesive, and relevant for theories of what you study.

A bit about the slides themselves. You want these to be slick and attractive. Aesthetics matter. Make your research look like it belongs in top-tier journals. For example, make your figures looks clean and clear like those in *Nature* or *Science* papers.

Many of the best talks include a graphical outline to show listeners where they are as they take the journey through an interesting bit of the scientific landscape. Animate the parts of each slide to come up only when you are talking about them. You want to control the attention of your listener. Don't present a wall of text and then read each line. You will lose the entire audience while they read and ignore you.

How should the talk end? Most people end with a list of future directions. I find this to be amateurish, like an undergraduate talk. Instead, tell them about the projects through the lens of the grants that will fund them. "My first R01 will use the methods I described to answer a new question about how memory supports more complex decision making in children. This will have three aims..." Then you can say that you are happy to tell them about other future projects in individual meetings.

Practice this talk, a lot. You want it to be smooth. You want to be relaxed and able to give a stellar talk even if you have not slept well for a couple of days. It is likely that you will be amped up enough for your first interviews that you will be a light sleeper leading up to your job talk.

If you are reading this early enough, start working on your talk a year or more before you need it. Sign up for a slot in your departmental seminar series and imagine that you need a job talk based on what you have now. Put that talk together and practice it until it is smooth and tight. You can also present your talk in a lab meeting. The truth is that in a year or two you will have a couple of more projects, but the meat of the talk will probably be essentially the same as it was when you were a senior graduate student or first-year postdoc. By starting early, you will enjoy significant savings when it is time to do it for real.

THE INTERVIEW

The interview starts well before you arrive. Specifically, you need to do your homework on the department ahead of time. You will get an itinerary beforehand and you should have looked at papers from all the people you will meet with. When I was really on my game, I would try to think of potential collaborations between myself and each person on my schedule. That way, I could tell each person about the experiment idea I had while reading their papers that we could do in my electrophysiology lab when I joined the faculty.

Your homework should also involve understanding the organizational units of the department. If they are hiring for the cognitive area, find out who the other people are in that area. Is there a clinical program? Are the faculty also affiliated with an interdisciplinary neuroscience program? With the existence of the internet there is no excuse for walking into an interview blind. If you do, this will communicate that you are not serious about coming to that school.

The other thing that you should have ready before you leave for an interview is a spreadsheet with your startup cost estimates. Why? Because some chairs will ask you for a ballpark figure of what you would need to start up your lab. If you haven't really worked the numbers, you are likely to give a low-ball estimate. Did you include two years of salary for a postdoc? How many data collection setups are included? I was caught off guard when asked this the first time during an interview, figuring that it would be a point of negotiation after getting an offer. The longer I was on the job market the more I learned that I didn't really need to have a precise answer, instead I could just tell them that it would be less than fMRI or two-photon imaging. But it takes a while to know how you fit into the world of startup packages, so it is best to know a round number, such as $250K, $500K, or $1 million. Also, if you don't know what you need, then you might not know whether the space that you are shown will be sufficient to house your equipment.

Okay, so avid readers of this book will have their suit from when they interviewed for graduate school. However, if that doesn't fit anymore or is out of style, then you need to go get a new one. For women, a nice pant suit with a couple of different scarves for the two days, and for men, a two-piece suit with two ties will get you through a two-day interview looking sharp. You will need to look nice both days. Often your meeting with the dean is not on the same day as your job talk, and deans are the most suit-wearing people you will meet. The faculty might not freak out if you wore jeans and no tie with your blazer, but most deans will think it is a sign that you are not serious. So look good. You can go back to hoodies and sneakers after you get the job.

You will need to ask questions of people. What is the teaching load? What is the procedure for making offers to graduate students? What is the sabbatical system? What is the tenure process? Is office space provided for students and postdocs? Can I buy myself out of teaching? What is the **indirect cost** rate (that is the percentage of funds that go to the university to support and administrate the grant)? How many credit subjects are in the pool, and does the supply run short of the need? This is just a sample of what you can ask the search committee chair and other faculty members. It would also be good to talk to the director of graduate studies about how priority is determined for graduate student recruiting, how offers are allocated, and whether they are capped.

Ask to see lab space of existing faculty to get an idea of what is possible. Ideally, you would also ask the chair to see the space that they are anticipating giving you. This will help you determine how your equipment would fit and where people would work. It will also give you a sense of how much work is needed to make the space ready. Typically, lab space is renovated before a new faculty member starts. I have seen big differences in how ready a department is for this renovation. At one end, some departments have the space cleared out and stripped down to the studs, waiting for you to work with the architect to design the empty space. At the other end, some departments have people still working in that space, with huge pieces of equipment that need to be moved, walls that need to be knocked out, and so forth. One of the terms you can negotiate is a deferral if it looks like the space cannot be made ready in a couple of months. And remember that a deferral is always a good idea if you are a postdoc, as we discussed in Chapter 5.

You should anticipate being asked some standard questions. The first and most obvious is, "Tell me about your first grant proposal." I call it the Randolph Blake question because I interviewed at Vanderbilt when Randolph was the acting chair and it was the first thing he uttered after I sat down. I found that at the top schools someone always asked this question and often it was asked by the chair, but not always. Generally, you want to know what funding agency and mechanism you would target (e.g., the National Eye Institute at NIH for an R01, or NSF at the Perception, Action, and Cognition program). This is something of a test itself because your answer indicates whether you know the landscape for funding and which funding agencies sponsor your type of work. You also want to be able to briefly describe the tasks and methods you would write into a grant, and how the experiments have a theoretical impact that could change what we think we know. The idea behind this question is to probe your level of preparedness for what comes next in your career.

Mature and well-prepared candidates will already be thinking about how to get grants even before their lab is set up. If this is you, you should know exactly what experiments you need to run as soon as data collection is possible to get the preliminary data into a proposal and demonstrate feasibility. As I mentioned above, it is a slick answer to say, "I will describe my first R01 at the end of my talk, but I have a number of other grant ideas," and then tell them about the other lines of work you imagine writing grants for and the types of agencies or foundations they would be well suited to target. Of course, you need to have thought about the aims for these grants. If you really have only one grant well thought out, then talk about that one and flesh out the proposal in terms of the pilot data you have already collected and how the questions and methods fit well with the goals of that agency. More grant ideas are not necessarily better if none of them can be described in any detail.

Other people will want to know what you can see yourself teaching. This is where your homework for the teaching statement pays off because you will be able to name classes that they have on the books that they need teachers for. Making up a new class is fine too; some people really like to hear how you could expand the offerings of the department. Keep in mind that proposing a new graduate class will have a smaller impact than proposing a new undergrad class, because enrollment for grad classes are pretty small, and graduate students do not generally pay tuition like undergraduates do. However, if you can offer a course that helps a department with its clinical APA accreditation, that can be a strong plus at the graduate level.

Another common line of questioning is whether you've thought about collaborations within the department, or across the university. This is a way of gauging your interest in the institution. Obviously, if you've done your homework on the people you are meeting with and can describe how you might collaborate with them in detail, you can give a strong answer. Having several potential collaborations that you can mention, however, is even better, because it shows that you have multiple points of intersection and would be a good institutional hire. As discussed in Chapter 9 in the section about getting tenure at a research institution, these collaborations are engaged in only after your independent lines of research are off the ground and paying off. You get tenure based on your independent papers. Collaborations are just gravy on top of the real meat and potatoes of your independent work.

Finally, people also have to live, send their kids to school, and have fun. It is a good idea to briefly survey the housing market wherever you interview. Asking the existing faculty what neighborhood they live in, or where they would recommend living to have a short commute, good schools, and fun local activities shows genuine interest in the place you where you are interviewing.

Keep in mind the task of the department interviewing you. They are trying to find someone with research that is interesting and important enough to get tenure, and with whom they could see themselves wanting to spend the next 5–40 years. Obviously, the job talk, CV, and application materials will mostly answer the question of whether the research will be productive and fundable. However, showing that you would be a nice person to have in the department requires you to demonstrate some social skills. You want to be yourself, but try to be your best self.

THE CHALK TALK

Some departments may ask you to give a chalk talk. This is a fairly unstructured question-and-answer period with faculty that can go by different names (e.g., round-table discussion). The goal is for the audience to ask theoretical questions that were not possible to probe in the short question period after a public job talk and to have you expand on your research plans for when you start your new position.

Most people do not plan any slides or materials for a chalk talk. As the name implies, you can draw on the board to show your thinking or to diagram the aims for your future grant proposals. However, I have seen people prepare additional slides , including a slide for each aim of their grant, with preliminary data, schematic diagrams of their recording equipment, and so forth. I have to say that it feels like cheating, but the slides very convincingly demonstrated the candidate's level of preparation, so I would recommend having them. To be polite, ask the group if it would be okay to show them.

NEGOTIATING THE OFFER

After the department has interviewed everyone, the search committee meets to determine a recommendation. Then there will be a department faculty meeting to decide to whom to make an offer. At these meetings the faculty often decides who is the first choice for an offer and whether the other candidates are above threshold for an offer if the first choice turns it down.

If they pick you, you'll typically hear from the chair of the department that an offer is being made, frequently by phone call, followed by an email that includes the official offer letter. This letter typical states that they would like to offer you the tenure-track position, the deadline for acceptance, but little else. Usually it states that a startup package will be negotiated between you and the department chair. Frequently the deadline to accept the offer is two weeks out.

This is when having prepared a spreadsheet with your startup funds request will come in very handy. You can send it along to the chair. They will then respond. Sometimes the response will be that the number you are asking for is fine. However, frequently the chair will seek to eliminate items from the list. For example, if you are asking for graduate student support, they might point out that support is possible through the college and so you don't have to worry about that. Or they might say that there is no precedent for postdoc salaries being included and that the dean won't approve it, so you need to exclude them. You are free to push back if you feel that any of the items are essential for your success. However, your options, if you get to a stalemate, are to decline the offer or accept it.

A couple of other things are included in these negotiations. One is the deferral that I mentioned above and in Chapter 5. A deferral is when you say that you want to accept the offer, but you want to start a semester later or maybe the following year. This is essentially negotiating the start date, as people routinely do in private industry. I encourage all my trainees to seek a deferral. Deferring has become the norm for people using neuroscience techniques. This is because they are usually finishing a postdoc. The other reason I encourage my people to defer is that their space is very unlikely to be ready for them if they are going to start in a few months (i.e., the coming academic year). Universities are great, but not the nimblest organizations in history. This means that the renovations of your lab space need to go out for bid, the department accepts one of the bids, and then the work needs to be scheduled and completed. If the building industry is fairly active in the area, it can be difficult to get this relative minor work scheduled. You will know this if you've ever had a house remodeled.

I know several people who were told that they could not get a deferral and instead, had to spend 4–9 months at their new job while their tenure clock was running before they even had lab space. After their lab space was ready, they still had to order and set up all of their equipment before they could start doing science. If your science uses nonstandard or customized equipment, it might take several months to manufacture and ship it. If possible, negotiate access to just enough startup money to order that equipment ahead of time, during your deferral. Then you can show up with your space ready and that essential equipment waiting in boxes. I hope that departments can get faster about preparing lab space so that I can give different advice, but until then, the deferral is one of the most important things to negotiate because it impacts your ability to have the strongest possible tenure package before your tenure clock is up. You want your tenure clock to actually begin after you have access to your lab space. If you do not get a deferral and have to wait to get into your space, lobby your chair to get time added to your tenure clock.

New faculty members often get their first semester off from teaching. This is fairly standard. The idea is that you can use this time to get your lab set up. If possible, ask to have your second semester off instead. If you don't get early access to equipment funds, let equipment vendors fulfill the orders and ship your stuff to arrive during that first semester. Then you can use the second semester to the fullest extent to get everything up and running. Also note that if the space is under construction, then having the first semester off is a particularly useless gift.

The second point of negotiations is salary. At large state schools, you can look up what your colleagues earn. The latitude of administrators might be very restricted. At all universities, there is going to be an effort to keep equity among faculty. If you ask for $200,000 USD per year as a first-year **assistant professor**, they are going to point out that this is a senior level salary and will almost certainly say no. Sometimes you can't really move them at all on salary. However, you have to ask for what you need to succeed. If you are moving with a family of four to a city with a high cost of living, such as New York, Boston, or San Francisco, then you need to make sure you can live on what you make. If you have to rent a smaller apartment than you had in grad school, you are unlikely to be happy and stay at that university very long. The school needs to know that. Again, however, you are in the position of declining the offer if your negotiation is unsuccessful.

Occasionally, the god of timing wills that you get a couple of offers at about the same time. This would give you substantially more leverage in negotiations. If University A is offering a better package, or part of the package is better, you can let University B know and see if they can match it or even beat it if you prefer to work at University B. But don't just do this to see how much you can get each department to come up with and play them off of one another. If you really know that you would rather be at University A, their offer is better, and they give you what you need to be successful, just take that offer. Don't spend weeks having the chairs scramble around going to the deans just to get this and that unless you are really on the fence about the two offers. You actually are doing University B a favor by turning their offer down quickly. They can then go back to their list and make another offer to someone else.

During the period between the offer and the deadline, many people make another trip up to the university where they interviewed to look for housing, to have a serious look at lab space, and to meet with people about hitting the ground running. Often the college pays for this as a recruiting cost. Get a realtor to show you apartments, condos, or houses, depending on your needs and financial resources. Again, don't do this just to communicate that you like the department before telling them no. That said, these trips can be very useful. During one such trip I found that I couldn't afford to live within 30 miles of the university. Another revealed that the lab space available would not fit the equipment I needed. Perhaps I should have figured these things out during the interview, but the schedules for the interviews can be packed and there is not always time to attend to these details when you have meetings scheduled back to back.

Please note that if you have a single offer, you still have options. If the university and department seem good and the offer has the basics of what you need, then you could go for it. But another option is to stay in your postdoc until something better comes along, assuming you have grant support to sustain you. If you continue to publish and work hard, something more appealing will probably come along. Another option is to take the job, recognizing that it is not your dream job, but that it does not need to be your forever job. If you get into your new faculty position and do well, you can always apply for your dream job when a slot opens up there. In fact, you might be a more attractive hire once you already have a tenure-track position and have demonstrated productivity with publications from your lab. This is because hiring someone who is under-placed and kicking butt is less risky than hiring someone who does not yet know how to successfully balance teaching, research, and service.

YOU'VE ACCEPTED AN OFFER, NOW WHAT?

Does it seem like time to party? That's because it is. You should celebrate. You have spent approximately a decade working toward this goal. You are about to embark on an extremely exciting time in which you will move to a new place, start your own lab, and work without the safety nets you had as a grad student and postdoc.

However, after the party has died down, there are a few things you can do right now to enable success. Prep the first class you are going to teach. Ask the person at your current institution who teaches the same class if you can have their syllabus and materials. Or ask the person you are probably taking over a class from at your new institution—often they are happy to share their materials and this will give you a head start. People are usually pretty willing to share this stuff. You could also ask your mentor or people at other schools who teach that class for their materials. You can also request desk copies of the possible textbooks from publishers and pick one out.

During this time, you should clear out the backlog of papers from your postdoc or grad school years. Although these are useful to keep your publication rate up during your first couple years of your faculty job, keep in mind that you soon won't have access to the equipment that you currently have in your mentor's lab. If projects need more data, collect those and focus on getting all of your projects wrapped up so that they only need to be written. Write up and submit experiments you can imagine reviewers may want more data for, so you can still be there to collect more data when you get the reviews back.

Finally, start sending emails to your friends in the field, letting them know that you have a faculty job, when you will start, and that they should send their best RAs, undergrads (who will be applying as graduate students), and potential postdocs your way. Recruiting high-quality personnel often takes longer than you think. Begin working to recruit people early. I've had some amazing postdocs, most of whom I started recruiting several years before they were done with graduate school. You are entering a stage in which you are both talent scout and scientist, so start scouting.

Ashleigh's interview with Lera Boroditsky, PhD

lera.ucsd.edu

Lera Boroditsky, PhD, is an associate professor of Cognitive Science at the University of California San Diego. She was previously a faculty member at Stanford and MIT. Lera and Geoff initially met when he was interviewing at MIT, supporting a piece of advice that Lera gives about viewing faculty interviews as exciting times to meet new people. Lera's research challenges notions about the relationship between thought and language. On the day I spoke with Lera I was in a cold house in Ohio while she sat outside in a California eucalyptus grove with a singing tree next to her. So can we all just agree that Lera has life figured out and you should treat her word as gospel?

ON TAKING THE PLUNGE

There is usually not a perfect time to apply for an R1 job, and Lera often advises people to apply in stages. For example, even if you don't feel completely ready, but a dream job or a few dream jobs become available, you should go ahead and submit an application because the cost of applying is not very high and you might regret not applying. If you are not selected, then it's no big deal. Indeed, when more jobs come available next year, you can apply, and you will be better prepared because you will have gone through preparing application materials once already.

Consider the opportunities for growth where you are if you are trying to decide whether to stay where you are versus moving on. For example, if you have funding and are doing great work, then that's a reason to continue in your current position. On the other hand, if you feel as though you need more resources or more independence to really get momentum, then it's a great time to find a faculty position that will give you resources and independence.

Lera added that almost nobody feels ready to apply for R1 faculty positions—you are always thinking about what you would want to get done first, such as publish more papers. You may always feel like your portfolio doesn't look as good as you want it to. But since it's a stochastic process, there is no harm in applying and giving it a try. You may even get some useful feedback from the application process.

INTERVIEWING THE INSTITUTION

People have different desires. If you're looking for an intellectual fit with several faculty already in your area doing similar work so you can closely collaborate, then during the interview process you should verify that such collaborations will be possible at the institution. However, some people favor working alone so it might not matter much to them who else is in the department from an intellectual standpoint. Ultimately, it is important to consider what you want from your colleagues and then find a department that provides what you want.

TWO-BODY PROBLEM (AND OPPORTUNITY)

A two-body problem is the academic term for when a significant other is also an academic both of you are trying to end up in the same location. Lera has dealt with two-body problems before. She said that having a significant other who is an academic presents both problems and opportunities.

It is certainly a more complicated process when there are differences in timing between the two careers. For example, often one person is going to graduate a little earlier or wrapping up a position or commitment a few years earlier than the other. Being on different time clocks can be extremely stressful if you are trying to precalculate where you want to be, which leads Lera to a few great nuggets of advice.

First, Lera points out that some states of being are temporary. If you are in graduate school or in a postdoc, it may be a worthwhile investment to live apart from your partner for a year or two if the result is a really great graduate career or postdoc position. Nobody wants to do it, but it's a temporary cost that makes a large investment in your future career—an investment that will pay dividends for the rest of your career.

Second, you ultimately have very little control over which universities and departments will be able to make accommodations. The only thing you really have control over is your own work. Your best chance of finding a two-body solution is if both of you independently do really great work. Great work will create more opportunities for both of you individually and therefore for both of you jointly. Couples with a two-body problem often feel a lot of anxiety around the uncertainty of the issue. Often people don't know if they will ever be able to solve it. Lera advises that you deal with that anxiety by doing your best intellectual work because that's really the only thing that will make your outcome better. All the other worrying doesn't actually change anything, but if both of you do excellent work, you will find a good solution.

How do you tell an institution that you have a two-body problem? Lera says that timing depends on the situation. There might be two jobs open at the same time and you both apply. There may be only one job posting and it fits one of you, but it may not be clear if there will be another job that fits the other partner. Lera advises that it's still really smart to apply even if there aren't two jobs, because universities do have some leverage in creating a second position. The university might really benefit from your having an awesome partner that they would want to hire anyway, which would give them the perfect reason to go to the dean and argue that they should hire both of you. In this case, Lera believes that a department could even see a two-body problem as a benefit.

Lera advises that you be smart about how you approach formally informing the department that you have a two-body problem, even though some of the people in the places you apply will already know you have a partner. This is because there are some things that can make it easier or harder for the university to help. If you've been completely secretive that you have a partner who is also going to need a job, and you've been made an offer, then it could be too late in the cycle. If you now reveal that you have a partner that also needs a job, then you've put the university at a disadvantage to be able to accommodate you. This is because at this point they may have already made a lot of commitments with other job offers and there just may not be any more money in the budget to help. So making your needs clear earlier in the process will give the university more flexibility and help them accommodate you.

A danger in telling the department earlier about your two-body problem is that it could create bias against you. The department may decide not to extend an offer because they precompute that you bring a difficult two-body problem which is not solvable. It is possible that they will decide not to interview you or make you an offer because they think they are wasting their time. You want to avoid that error as well. Ultimately, Lera believes that it is best to be open from the very beginning. Especially at large state schools, where there will be a long process and they may not be able to turn it around in time if you don't tell them early enough.

Flexibility is really important because it may be that you don't find the perfect solution at any one university. There may be a split position at one place or a different kind of position at another place. Be creative in trying to solve the problem by looking at other universities nearby. Look for a position that might be converted into a different position later. For example, it may be that this year there is only a lecturer position available for one of you, but there is going to be a better position coming up in a few years that one of you would be a good fit for. Think creatively in terms of location, time, and position type.

It is inevitable that at some places one of you will be recruited more strongly than the other. The reverse might happen in another place. There is this phenomenon of one of you being the "yucky spouse," as Lera says. To psychologically deal with this, Lera points out that these things are very temporary and arbitrarily contextual depending on what jobs happen to be available and who happens to be running the job search. As much as possible, try to ignore the history of how you were hired. When you arrive, engage with your colleagues as if none of that happened. Lera refers to the metaphor that faculty hiring is like sausage making: the less you know about the process, the happier you are with the results. When you do arrive, it's best to forget the process by which you got there. This will be better for your relationship with your significant other and better for your relationship with your colleagues. Start fresh and treat each other and everyone else as full colleagues.

MENTORING STATEMENT

Lera agrees with Geoff that it's a really good idea to add a mentoring statement to the teaching statement, particularly if you haven't had a lot of experience and are applying straight out of graduate school. It's good to show you are engaging with mentoring and teaching thoughtfully. Before you get to writing the plan, it's best to actually have experience mentoring. To that end, recruit undergrads to work with and really mentor them. In addition, do some peer mentoring. Almost every lab will have trainees who come in after you and you can mentor them. Teach them how to conduct research in the lab and collaborate with them. Take advantage of all opportunities to do peer teaching and peer mentoring. Write down your philosophy of how you want to structure your lab. Do you want to structure it in terms of one big project that everyone contributes to? Do you want to structure it with individual projects for each student? Why does that make sense for the kind of work that you do? This is all useful to think about ahead of time anyway since you'll have to implement it eventually. Consider different models and explain why you like a particular model. Not only is it a helpful exercise for you, but it will also show your search committee that you've thought about it.

TEACHING

Lera has never thought there was a disconnect between being a good researcher and being a good teacher. Being a good researcher means you have the ability to think about things clearly and explain complicated topics effectively to your colleagues. That's what it means to give a good talk at a conference and to communicate your work. Being a good teacher is this same ability applied to a different context. Teaching and research share a lot of skills in terms of being able to lay out the logical structure of a topic and explain complex information.

Lera advises that you should get teaching experience when you are in graduate school. You should TA an undergrad course or teach a summer school class. It's one thing to take a class as a student and learn the material. But when you TA a class, you learn how to teach that material. You learn the approaches the professor takes to explain the material, what works, what doesn't, what are good demonstrations, how to control a classroom, and what kind of engagements for discussion work and don't work. Lera advises applicants to definitely get such experience before their first faculty position because not having it will make your job a lot harder. Finally, mistakes are more costly later, when you are on the tenure track. If you are a graduate student TA and you don't plan your lecture well, the stakes are much lower, but if it's your own class and you give a bad lecture, there are more consequences.

Ultimately your job talk is taken as a cue for what kind of instructor you will be in the classroom. If you are clear, persuasive, and you answer questions well, it is very easy to imagine you being successful in the classroom. Because those are required teaching skills, it's easy to judge whether you are able to take on that role. Certainly, some experience and some teaching evaluations from classes you've TA'ed also provide supporting evidence of your skillset.

RESPONDING TO QUESTIONS DURING THE JOB TALK

Lera said that people underprepare for answering questions that arise during their talks. That's when the faculty really see what it will be like to have you as a colleague in the department. The chance to engage with a question is where you can really shine. Do you take the question seriously? Are you able to think about it, meet them halfway, engage with them where they are in terms of their expertise and thought process? What will it actually be like to discuss things with you? Her recommendation is to practice answering questions. One way to do this is to try your talk out on really broad audiences so you get a broad set of questions and you'll be ready for a lot of them. How you answer questions is an important part of how your colleagues evaluate you.

LERA'S OTHER JOB TALK TIPS

- Practice humility; if you don't understand a question, do your best to get to the bottom of it.
- Learn when to say, "You know what, I'd love to understand this question, but right now I'd like to tell you about our last few experiments. I would love to talk to you afterwards."
- Learn how to be open and think on your feet, how to embrace questions.
- Don't be too humble; don't agree to criticisms you don't understand or just acquiesce to criticisms unless they are really fair. You want to come to a fair representation of what research has been done previously and what you have found in a thoughtful and calm manner.
- Remember that you've been thinking about this stuff longer than anyone else. Remember that you are the expert.
- Don't just passively accept all criticisms or comments. That doesn't come across well because some people in the audience won't be able to tell that you actually have an answer.
- Don't be too rigid. Admit weaknesses if they exist. Talk about what you did and want to do to address weaknesses. If someone has a good criticism and you haven't yet done something to address it, show him or her that you really want to by saying what you would do. For example, say, "This is something I also worry about and I really want to address it as well. What I really want to do is X and Y, and if that turns out, then Z." This allows you to think together with the person who asked the question, which is an effective way of engaging, rather than being on opposing teams.
- Don't dismiss _anyone_. Sometimes people dismiss a question because it comes from a person who says something kind of funny or a young person who doesn't look important. You don't know who those people are and the people in the room do, so take every question as a valid intellectual engagement and do your best to engage with it. The people in the room are going to decide if the question is serious or not, not you, so don't dismiss someone because they don't look like the kind of person who would be important. This would be a terrible mistake.

BE READY TO ENGAGE EVERYONE

Lera reiterated Geoff's point that you should learn about the department that you're interviewing with. She has seen people who don't know much about the department or what the faculty members do when they arrive for their interview.

When you walk into an interview you should have some sense of everyone's research. Of course, there are some people whose research is close to yours and you should know their research really well. There will be other people whose research is quite different from your own, but at the very least you should read one of their papers and a few abstracts. Have some questions about their research and be ready to engage them. If you can find a point of connection, usually you can ask them a question that would be helpful in your research. People love that. Take the opportunity to make them the expert and draw on their expertise for something that you're genuinely curious about. That's a role that people love to be put into.

Young interviewing faculty candidates may not feel comfortable engaging with graduate students because a younger candidate might think, I'm still a student myself so I'm going to be buddies with them. That sometimes means they don't intellectually engage the students when they meet with them. This is a mistake because the students also want to talk about their research. Ask the students about their research and engage them in the same way you would a faculty member. In one instance, an interviewee spent their lunch with graduate students asking if there were any scandals in the department. The students hated that because they are serious researchers and wanted to talk about work. Ultimately, even though the faculty really liked this candidate, the feedback from the graduate students actually prevented this person from getting a job offer. Always be a professional intellectual!

Be excited to be in the place. Think proactively about how you would fit into the department. People will ask, "How do you see yourself fitting in here?" This is especially the case if it's a department that is a little off-center from what you do. It is important to describe the connections that you see and how they will benefit your research. You also want to illustrate how having you around will be beneficial to the environment; for example, how you will enrich the program and how your research will push the department into new directions. Often young applicants are so excited to be applying for a job that they might not do this kind of lobbying well. Lera points out that departments are not trying to reward you. What they are trying to do is to bring in a good colleague that will reward them. What are the ways that the tools you are bringing will enrich the process of educating graduate students and possibilities for collaboration? What expertise do you bring in terms of teaching? Will you be able to add potentially large, interesting courses or release senior faculty from some of their teaching obligations? Think about the ways in which you are going to make their lives better, not just how awesome it will be for you.

FRAMING CONCERNS

If you have a particular concern about an institution, ask questions related to that concern and try to put a very positive spin on it. It is better not to bring things up as concerns, but rather to explore and learn as much as possible about the topics that you're interested in. So as much as possible, avoid framing them as concerns. Try to inquire about the proactive things you need to do to succeed. People do recognize when there is some trepidation or negativity in your questions and think you are not excited about working with them.

In general, the positive, proactive framing is also how you should ask questions of colleagues. If they ask, "Do you have questions of us?" It is useful to ask them, "What do you think is a good path to success here? What would it mean for me to be a good colleague, a good member of your department, a successful member of your community?" And then hear what they say because it will reveal a lot about the culture of the department and what is valued there.

The question of what makes a good, successful colleague in their department is a question that you should ask multiple people. The department is not a homogeneous mass. In it there are different people with different opinions. Your success will depend on the opinions of many people. There will be some who will be your closer colleagues and you may want to weight their opinions more strongly, but it's always good to hear the broad perspective.

CONFIDENCE

Until you do the job you don't know that you can do that job. There are many different paths to success and many different environments. You may be very successful in one type of environment and less so in another kind of environment. In general, do all the things you can to find out for yourself if you like all the components of an R1 job.

Leading a research program and mentoring undergraduates shows that you have intellectual leadership in your project. During your graduate and postdoctoral career, do the kind of work that will let your letter writers say, "This person took intellectual leadership and they will have no trouble formulating a research plan moving forward." Be ready to tell people what your research trajectory will be over the next five years. People will ask a young applicant whether they have thought beyond their dissertation. Do you actually have a plan for what you want to do? Being prepared for this echoes Geoff's advice about knowing what your first grant will be about. That gives the search committee information about whether you are good at planning and designing a line of research, but it also gives you information about whether you like doing those things.

Get experience writing grants before applying for R1 faculty jobs, if you can. At the very least apply for fellowships. Try to help your advisors formulate grants. Learn something about the grant application process because if you haven't written any grants yet, that will be a new aspect of your job that is very different from being a graduate student. No one ever feels like they've done everything perfectly in their careers—that moment never comes—there are always many things you can do better. But be excited for the project and the different parts of it. People will respond to that excitement and they will want to have you around.

NEGOTIATION

I can very much identify with Lera's background. She was raised with the view that money is terrible and everything about money is shameful. So although it always goes against her upbringing to enter the negotiation process, Lera had some great nuggets of advice regarding negotiating your hiring package.

When it comes to salary, remember that you are entering a university where you are going to get incremental percentage raises from your starting salary and that cumulative effect makes that base pay really important. For example, a 3% raise on a starting salary of $50,000 USD per year is going to be very different over time than a 3% raise on a starting salary of $100,000 USD per year. Another reason salary matters is that you may choose to take on more teaching or an administrative position for more money, but that will take away from your research.

Don't be afraid to ask for what you want, but always be nice about it. It's useful to have comparators. If you have a number in mind, be prepared to show that similar people in similar positions have that salary. For state schools that information is public. Most of the time you'll be successful negotiating if you have other offers from other universities. You can also make arguments based on cost of living. For example, the same salary is not actually the same salary in Montana and Manhattan, so you should include a multiplier for the cost of living.

Salary isn't the most important thing that you negotiate because there are lots of other things that matter for quality of life in a department. Lera suggests treating the negotiation as a complete package rather than piecemeal. Think about what is most important so you can focus on getting some of those things. These might not be financial at all. For example, it might be teaching relief early, or teaching assignments later on.

At each individual institution, there will be flexibility on some things and not others. You want to have as broad a sense as possible of what you are seeking so that you can leverage where the institution is flexible. For example, some universities will have no flexibility on teaching load, but they will have flexibility on TA funding, which can reduce the burden of the teaching load. Or, they may have no flexibility on salary, but they might be able to offer support for daycare for your kids, or an extra benefit for schooling, or an extra benefit for buying a house. All those things are just money in the end. You should think about a certain standard of living that you want to achieve and the multiple routes to achieve that. Maybe the flexibility to achieve that won't come in salary, but will come in another benefit, which might end up being more valuable than salary.

When you are negotiating a package, you will generally be talking to the chair of the department. But they often aren't the person controlling the resources. Some people feel shy about negotiating because they feel like they are taking resources from the department for themselves. However, these are typically resources from the dean. Therefore, you are trying to provide a really solid set of arguments to the chair to liberate money from the dean to the department. It's good to find out your situation: is the department chair an advocate for you, or are they the person actually controlling the money? If you are in the former position, use the chair as an advocate to bring resources to the department. This is also a way to start building connections with colleagues. For example, there may be something you need for your startup that would benefit other people in the department (e.g., an eye tracker). Because it would benefit other people as well, you may be able to get a nicer one or more money allocated for it. To figure out whom you are really negotiating with, just ask the chair. Is there a standard amount that the dean has already allocated, or is there flexibility for negotiating with the dean on an individual basis? Some departments just get a budget and they have to make their new hires work within that budget. Other departments will serve as a negotiator between the dean and each new faculty member.

CONCLUSION

Everyone sees applying for jobs as a very stressful time because it's not very nice to know that you're being evaluated. But it can be a whole lot of fun. It's an incredible opportunity to have engaging conversations with many interesting people in your field. Even if you don't get the job, you may make a good impression and end up with friends or colleagues for life. So treat this as an opportunity to meet people with whom you are going to interact for many years. Don't make this process transactional. Just treat them with interest and respect as real colleagues. When you do, they will treat you that way too.

Have the long view in mind. If it's not the right time for you to be hired now, for whatever reason, you're still making an important impression and creating an important relationship. Make sure you take advantage of the opportunity and enjoy the process of meeting fellow scientists.

CHAPTER 8

How to Succeed in Your New Teaching Job and Get Tenure

by Ashleigh

It is unlikely that your first teaching experience is at your first full-time teaching job. This is because securing a tenure-track teaching job, and often even non-tenure-track teaching jobs, such as a lecturer position, usually requires a good amount of teaching experience, either by working as an adjunct, guest lecturer, or as a TA. Given that you likely have teaching experience, in this chapter I address some issues you might encounter for the first time now that you have landed a full-time teaching job. The topics I discuss in this chapter broadly apply to your first full-time teaching position, in both tenure track and non-tenure-track appointments, with the exception of the final section on tenure. Recognizing that there are a wide variety of institutions that fall into the umbrella of teaching schools with similarly varied expectations, here I focus on personal experience.

What follows are issues that have come up for me or close colleagues as we navigated our first tenure-track teaching job and the relatively modern approach to complementing the typical full-time teaching career with a generative research program. The topics covered here might seem surprising for individuals who have not begun working in a faculty position. New faculty are often surprised by the type and scope of decisions that have to be made during their first few months or semesters. These decisions range from how to spend your time meeting departmental teaching demands, research, and service, to what type of relationships to cultivate with faculty and students. If you are unaware of what this type of position might entail, you may not appreciate these new responsibilities or the requirement that you independently navigate these issues without your advisor to follow as a role model. I hope that this chapter serves as a guide if you feel alone at this juncture, or a conversation starter if you have a network of colleagues with whom you can navigate your new teaching position.

The second half of the chapter includes my interview with Provost and Vice President of Academic Affairs, Glenn Sharfman. His interview provides a different viewpoint from mine on succeeding at a teaching institution. His is the view of someone who has spent his career cultivating excellence in hundreds of professors at a variety of liberal arts institutions.

THE CANDY BOWL

Throughout this chapter, I am going to use the professor who keeps a candy bowl in their office for students as an example of what you do not have to do. However, before I do so, I want to clarify that some of my best friends have candy bowls in their offices. Some of these friends are classic *candy-bowl professors* in that they frequently welcome students into their homes, providing them meals and light when electricity goes out on campus, volunteering to run the food truck at the women's basketball team fundraisers, and spending countless hours supporting students through all sorts of personal issues, especially first-year students who are struggling with being away from home for the first time; and I respect and admire my candy-bowl professor friends immensely for doing so. My negative reaction to the candy bowl is rooted in the fact that I have never been a candy-bowl type of professor and it took me years to realize it was ok not to be. For me, the candy bowl is a metaphor for a number of traits historically expected of professors at teaching institutions (e.g., constant accessibility, blurring of personal and professional lives in an informal relationship with students, electing to talk to students rather than work on research) that doesn't have to hold true in order to serve students. It took me a long time to recognize that I did not have to be a candy-bowl professor to be an effective faculty member at a teaching institution.

When I began my first tenure-track teaching job I was 28 years old. My colleagues, who were old enough to be my parents, were incredibly nurturing toward the students. In return, the students adored them, looking at them like the cool parents everyone wants but few people really have. I watched my colleagues spend hours mentoring students, inviting them into their homes, attending their sporting events, cooking them Sunday dinners, staying on campus late at night for club meetings, and so on. The model that I saw for how to be a great professor didn't seem like it was going to work for me for a variety of reasons. I felt like I didn't have enough life experience to dispense advice, so I didn't want students to come to me with their problems. I also had my own three young kids at home to raise, so my hands were too full to feel like a parent to my students too, and I could never make it to their sporting events or stay on campus for late club meetings because I was a single mom of three. I wanted to work on my research and didn't see how I could be productive in the lab if I allowed the students to have constant access to me. Finally, I was trying to establish a professional relationship with the students, which can be a struggle for a young, casual woman, so I needed to not feel like their friend. It took some years to feel like it was acceptable to not want to be maternal to my students, and I settled for having the department reputation of being the older sister who did not want to talk about feelings.

The trajectory of my life was altered multiple times by professors who sat with me for much longer than was required and gave me tough, honest advice. I'm not saying you shouldn't do the same for your students and I am eternally grateful some faculty members are the candy-bowl type. I am encouraging you to figure out whether or not you are the candy-bowl type, and not to let anyone make you feel badly about whatever feels most natural for you. The advice in this chapter is primarily what I think those of us who are not natural candy-bowl professors need to know and would have really helped me to know when I embarked on my first teaching job. If this chapter were only about tenure, it would be organized into three sections on teaching, research, and service. However, because those expectations really differ for every institution, I instead discuss topics I could have used more guidance on when I was in my first few years of a full-time teaching gig. Then I include some sample questions to ask your department chair, dean, or tenure committee so you can get a better pulse on your specific institution's tenure expectations and process. Finally, my interview with Glenn Sharfman gives you one dean's perspective on a number of issues I raise and some of his own.

TEACHING EVALUATIONS

Teaching evaluations can be heavily weighted in consideration of your contract renewal each year, and ultimately, tenure if you are on the tenure track. Thus, it is easy to fall prey to an obsession with teaching evaluations and in turn teaching in a manner that you think will garner positive evaluations at the cost of helping your students achieve their highest potential. I aim to arm you with a perspective on your role as an instructor that will help boost your immunity to the lure of winning the teaching evaluation popularity contest.

Here is my favorite analogy regarding teaching evaluations. When we used to put our young children in timeout for making a bad choice, we didn't expect them to immediately thank us for it. This is despite the fact that we were putting them in timeout for their own good—usually either to calm them down so they wouldn't hurt themselves, or to help them learn how to treat other people so they could continue forging healthy relationships. These were noble, important, responsible lessons we were trying to teach the kids, but we weren't so crazy as to think they were going to realize the importance of the lesson as soon as the timeout timer went off and they could go back to playing. I believe that having students fill out useful teaching evaluations at the end of the semester is akin to expecting a child to thank you immediately after they get out of timeout. The reality is that at the end of the semester, students don't know whether you have helped them reach their goals. Students know what grade they earned and how hard they worked, but they don't know much else in terms of how you impacted their ability to achieve their future goals.

My approach to teaching is to remember that students are in college as a stepping-stone to their next goal, such as going to graduate school to become a child psychologist, or a research scientist in cognitive neuroscience, or other future endeavors that may not directly involve psychology. Treating the college degree like it's the end goal undermines and underestimates your students' capabilities. Taking the perspective that the reason students are in college is to secure a career in the long run, and not just a degree, sheds light on how ridiculous it is to judge the effectiveness of your teaching by administering surveys to your students at the end of the semester. The seeds that you plant to help your students get into graduate school by providing them with a better understanding of research, or improving their interviewing skills with feedback on their oral presentations, or teaching them how to read journal articles so that they will excel in their graduate school classes, will not bloom by the end of the semester. If they don't go to graduate school, they've still learned how to deal with difficult material, how to work through frustration, and how to articulate their thoughts. The quote by Joseph Pearce, "All seeds contain life, but those dropped among rocks and thorns simply stand far less chance of expressing that life than those falling on fertile soil," helps emphasize the long view of your role in the classroom, and the absurdity of measuring your success with end-of-semester teaching evaluations. I believe that it would make considerably more sense to judge your worth as an instructor by student feedback solicited years down the road. We all have those professors who we look back on fondly years later, when we realize they set the bar high and taught us how tall we could stretch by taking the time to write extensive, thoughtful, and encouraging feedback on our assignments, or taught an introductory course with such passion that we decided to major in the topic and take all their classes. It is this long view that I think is the true measure of good teaching.

Recently there has been a growing interest in conducting empirical studies to examine what teaching evaluations actually measure. Unsurprisingly, the outcome of these studies does not support the use of traditional institution-wide teaching evaluations. Concerns with teaching evaluations include sampling rate; that is, less than half of the students typically complete them. When a low rate of students complete evaluations, you are likely drawing from those who are motivated to complete them, usually because they are either very happy or very unhappy with their experience.

Another concern with teaching evaluations is the extent to which they become a popularity contest. Often the professors who are highly rated are the ones who bake cookies for students on evaluation day and have candy bowls in their offices. In my view, a professor who has a constantly open door to students may not be spending much time on other tasks that support students, but require time away them, such as research and manuscript writing, which I discuss below. This willingness on the part of a faculty member to be in so much contact with students outside of class also leaves the faculty member vulnerable to potentially inappropriate conversations, like being asked what they know about other faculty members' private lives. Below I discuss the close relationships students may expect to form with their faculty members as another issue, which is unique to some teaching schools and can be difficult to navigate.

A third reason that there is concern about teaching evaluations is gender bias. A growing literature on the topic of gender bias shows that female instructors get lower ratings than men. In my own informal polling, students have found it acceptable to comment on what their female instructors wear to class, but not what the men wear. This is supported by a growing literature on the topic.

Fourth, teaching evaluations do not occur in a vacuum, but instead, can reflect the larger context of the semester. For example, I have a friend who was teaching environmental science and got particularly low evaluations the semester following the presidential election of republican candidate Donald Trump. She knew she had a particularly conservative group of students in her class, but approached her institution's center for teaching, seeking feedback and advice on improving her teaching. The representative at the center for teaching told her that across the university, classes that dealt with the environment, race, class, and gender received particularly low evaluations the semester following the election.

The multiple problems that accompany teaching evaluations are made more problematic by the fact that student teaching evaluations are typically the bulk of the feedback that instructors receive. Student teaching evaluations may in fact be the only feedback received, other the dean or department chair potentially evaluating your teaching once a year, although some institutions also require peer feedback and teaching reflections in annual reports or tenure dossiers.

I describe these issues with teaching evaluations to illustrate the necessity to arm yourself with some techniques that will provide you with alternative, potentially more accurate measures of your teaching effectiveness as you begin your career. This chapter's goal is to help the non-candy-bowl professor conceptualize what success at a teaching job looks like. Regarding teaching evaluations, invite your colleagues to come to your class and give you a written teaching evaluation.[1] Most of us should be doing this more often than we do. The semester gets started and things get busy and you may forget to ask your peers to come to your class. Before the semester begins, as you prepare your syllabus, send out some requests to colleagues to attend your class and evaluate your teaching. Typically, you should return the favor by going to their classes and providing a written peer teaching evaluation for them. These optional peer teaching evaluations provide written documentation of your teaching should you find a need or desire to submit them with your annual reports or your tenure dossier. If they are particularly negative, you can simply incorporate changes bases on their comments into your future lectures and toss the evaluation. Since you are inviting them, you don't have to ever show them to anyone if they are not the glowing evaluations you were hoping for, but you can definitely still learn from their feedback.

I also want to make very clear that there are times when student evaluations of teachers are useful. For example, informal mid-semester student feedback in response to very specific questions tailored to your course, or to concerns you have about student performance, can be incredibly useful. Direct questions about the organization of the course or the availability of the instructor often elicit more useful responses than questions about

---

1   Use the excellent rubric created by Hill, J. C., Thompson, C. A., & Beers, M. J. (2018). Classroom observation rubric for novice instructors. Poster presented at BISTOPS: Biennial International Seminar on the Teaching of Psychological Science.

the overall value of the course. Sometimes students can be perceptive about things we don't realize, or tend to take for granted, so this technique might lead to better and more helpful comments. I know that if I look back over my ten-plus years of teaching, of course my teaching has been improved by addressing issues that have been raised in student evaluations.

TIME MANAGEMENT

Above, I described some issues with teaching evaluations and that it is easy to become obsessed with judging your effectiveness in the classroom (and thereby your effectiveness at your job at a teaching institution) by teaching evaluations. Next, I want to help pull you back from that edge, primarily to help you understand that you can be happy at a teaching school and feel a great sense of belonging, accomplishment, and pride in a teaching job, even if you aren't a candy-bowl professor. The advice herein will also help you stay marketable should you need or want to change jobs, and provide you with talking points for your annual report and tenure dossier, about how you serve your students by working on tasks that sometimes take you away from students and require that you close your office door to focus on other tasks.

Returning to the perspective that your primary goal in teaching students should be to help them get to their next step, in psychology a sizable portion of that is the letter of recommendation or reference you provide for them. Specifically, the more impressive you are as a member of your academic field, the more impressive your letter of recommendation will be, and the more you can help students get into their preferred graduate program or land the job of their choice. In order to be an impressive letter writer, you should run your lab, program experiments, collect data (or have students collect data for you), analyze data, write papers and submit them for publication, travel to and present at conferences, network by giving talks at other institutions, peer-review colleagues work, serve on editorial boards for journals in your field, apply for grants, and more! Many of these tasks take you out of the classroom, away from your students, and require focused attention. Attempting to create space and time for your research can appear unpopular at institutions that tout close and frequent contact between faculty and students. I find that explaining to students what I am doing when I work off campus or close my office door, and why I am doing it (e.g., to write grants or serve on an editorial board), makes them very understanding and respectful because they look forward to my increased visibility in the field being an asset when they ask me for a letter of recommendation down the road. It is also helping me become a better teacher—because I get practice explaining topics through grant and manuscript writing and I have 'teaching moments' with my students where I explain my job and the logic behind my time management to them. One way to carve out this time if your institution asks you to post your schedule on your office door would be to block out hours, or maybe even one day per week, as a writing day, with a little note that says that you are writing letters of recommendation for students, grants that will bring funding to your department and research experiences for your students, and manuscripts that when published, potentially co-authored by students, will promote the institution. Another idea is to spend some time on the first day of class each semester explaining that the gravitas of the students' degree is only as good as the gravitas of the institution and faculty, and that you promote your university or college by publishing and presenting at conferences, visibly listing your affiliation on publications, posters, and presentations. Often students simply don't know what faculty do and why they do it. I find that students really appreciate being clued in to who you are, what you do, and why you do it.

There are numerous time-management apps that help people schedule an hour of writing a day or shut down background applications. One of my favorite nuggets of time-management advice is to check your email once or twice a day and not on weekends. I am not perfect at following this advice, but it's amazing how many problems resolve themselves and how many people stop wasting your time with emails when you become just slightly slower at responding. I know that it is important to be responsive to your students and colleagues, but if you let them know about your email responding schedule, they will likely be respectful of your boundaries and appreciative of your transparency. I suggest explaining in your syllabus that you check email once a day and not on weekends. I also advise letting your students know how long is an appropriate amount of time in which to expect a response and what to do if you haven't responded. For example, tell students that expecting a response to their emails within 24 hours during the week is reasonable and if they haven't heard from you, to please resend it with something catchy in the subject line. I do this with my students and it leads to some great email subject lines like *FLYING ALPACAS!!!* Or one that had five avocado emoji's in the subject line before the name of the course. I have also started telling students to only email me through the course website. I set the course website to keep all the email and not forward it to my computer or push it to my phone. I tell the class I will check it once a day and not on weekends. Since it's not even on my laptop or pushed to my phone, this has been very successful in that even if I do check my email, I don't get overwhelmed and distracted by my students' emails (but I still get to them in a reasonable amount of time). This method ensures that all my students' emails get the attention they deserve because they are all in one place, and it also keeps my inbox much clearer so I don't overlook communication about my scholarly activity.

YOUR RESEARCH PROTECTS AND PROMOTES YOU

Next, I want to discuss why you should maintain a research program and continue conducting research even if it is not a job requirement. The argument for continuing to run a lab and be active in your field can be divided into three main reasons: to benefit your students, your institution, and you. Above I discussed why running a lab serves your students. Now I want to describe how maintaining visibility in your field through research activity serves you.

First, the happiest employees tend to be the ones who can leave their job for another position if they so choose. Deciding to stay in a job, rather than feeling stuck there, drastically contributes to job (and life) satisfaction. This means that you should work to stay marketable even if you plan on staying at your current institution. It is easy to settle into a teaching job and spend most of your time teaching and in service positions because at many teaching institutions you can earn tenure even if you become complacent about your scholarly activity. The downside of this is that, in today's market, it will be very difficult to get hired at another institution if you've been exclusively teaching for years. This is even true if your new job is also at a teaching institution. Why is this a problem? Even if you think you have landed your dream job, it makes a great deal of sense to maintain your marketability because what the future holds for your personal life is always unknown. I often advise our students to try to keep their options open when it comes to picking a major or graduate school, and that advice also applies to you in your career.

The awareness that life can bring about the unexpected suggests my second main, self-serving reason you should maintain your research agenda: even though you may have landed your dream job in your dream city, life happens. I know people who have needed to move for all kind of reasons, including to be near a dying parent, to be closer to their new spouse, due to custody situations, to be close to a hospital that houses the world expert who can treat a specific ailment of their newborn child, because their institution does not offer maternity leave and they need to live closer to family, and because their new job is not the great fit or the happy place that they initially thought it would be. An institutional issue that might have you looking for a new teaching job is that small private colleges without large endowments have seen increasing financial pressure and are in increasing danger of downsizing, or in extreme cases, closing altogether. The list of reasons that you may not stay in your first (or current) job is endless, largely unforeseeable, and not necessarily an indictment of the institution, the administration, or you. You will want to have career mobility in the face of unexpected changes in your life circumstances. In sum, if you don't care at all about your students (I am joking, I know you do), then for the sake of being happy and marketable if unexpected life events occur, keep running your lab at your new teaching job.

AUTHORSHIP ISSUES

I discussed authorship issues in Chapter 2 to encourage undergraduate students to understand what they need to do to earn authorship on papers and conference presentations. As a faculty member, it is really tempting to include student authors on your work for a variety of reasons. For one, becoming senior author on a paper is an important step in your career. Second, it will help your students get into graduate school if they are co-authors on publications. Most relevant to this chapter though, is that it might help your case for tenure.

I caution you against the allure of allowing your undergraduate students to author papers unless they are ethically deserving of authorship. What specifically warrants authorship and what the order of authors on a manuscript conveys can vary from field to field (e.g., in our field, last author typically indicates the senior advisor and first author usually indicates the person who wrote most of the paper).

To provide some reasonable guidance in line with our field, I suggest using the rule of thumb that every author must provide significant intellectual contribution to the project. In my field, that means that simply collecting data does not warrant authorship. If you have a student or research assistant working for you who earns course credit or money for running subjects, they are already getting reimbursement for their time in the form of course credit or money. If you want to include student authors on your paper, help them intellectually contribute to the project, possibly by including them in discussions about the project (e.g., where it should go next, whether the current design makes sense or is optimal) or including them in writing the paper (e.g., helping with the literature review, editing).

In addition to who should be an author on a paper, you should also discuss author order with everyone involved on the project. Author order (i.e., the order authors' names appear in on the paper) in my field of expertise typically goes as follows: usually the head of the lab will take last author (i.e., be last in the ordered list of authors) and serve as corresponding author; the corresponding author comes with the serious responsibility of checking the paper proofs and reporting grant support. The first author is usually the author who conceptualized the experiments, programmed the experiments, and wrote the first draft of the paper. In labs at research-focused institutions, the person who fulfills those first author roles may be a graduate student early in the graduate student's career. Only much later in the graduate student's career, such as in their fifth year, might they become corresponding author. If you are running a lab at a teaching-focused institution with undergraduates, it will almost always be appropriate that you are both the first author and corresponding author because based on your training and the undergraduates' lack of expertise, you are more than likely contributing the most to the project in terms of conceptualizing the project, programming, and writing. If you are first author and the remaining authors are undergraduates, the second author should have contributed the second most to the project, and so on.

It's also very wise to have open, ongoing conversations about authorship issues when you start a project or when you start working with a student. Preferably, have this conversation documented in the form of an email or a lab meeting agenda to help prevent any confusion. I very much believe that you can run a lab with only undergrads and be very productive (see Chapter 11 for how to run a lab without graduate students). I also encourage supporting undergraduate students in earning authorship and mentoring them into graduate programs. However, you should be cautious and take ethical considerations into account, revising your policies and communication regarding authorship issues as necessary.

ESTABLISHING BOUNDARIES

A teaching job can be all consuming in many ways that are different from the way a research-focused faculty position is. One reason is because teaching schools often market themselves as places where students can get to know their professors outside of the classroom. Although this varies by institution, at some schools it is not uncommon for students to go to their professors' homes for meetings, club gatherings, or celebrations. Students at teaching schools may become more involved in your personal life in ways that you would not expect if you did not attend one of these institutions as a student yourself. The reciprocal relationship is also true, in that you might become more involved in your students' personal lives than you expect or necessarily desire.

The close-knit nature of small schools applies not only to relationships with students, but also to other faculty members. It is possible that someone will suggest or expect you to socialize with your colleagues outside of work. Of course, you have every right to lead your personal life in any way that seems like a good fit, natural, and comfortable for you, but don't surprised if you have to enforce boundaries that feel natural and right for your own life. These are important issues to think through before you have to confront them.

These boundary issues require determining how to balance and separate work and life experiences. The degree to which you divide work and personal life will be largely based on your own desires, experiences, and expectations. In the final chapter of this book, I discuss tips for so-called work-life balance, but that issue is separate from than the topic of how far you are willing to let your colleagues and students into your personal life and how involved you want to be in their personal lives. You might consider soliciting advice on boundaries from your senior colleagues or inquire about it during your interview. Having your own boundaries that feel good and reasonable is important to at least think about before you begin your new job.

When faculty socialize with students, it blurs the line between the mentor/teacher and the student. Blurring the line at best impairs objectivity between the faculty and students and at worst has disastrous effects on one or both parties, their careers, and their mental health. Some faculty can handle blurry lines relatively well and others cannot. That's not an indictment of anyone, but it's useful to contemplate where you stand on this spectrum. It's also useful to consider the specific circumstance. For example, going to a faculty member's house for a picnic with a large group of students is very different than one student going alone. I think that most people would agree that drinking alcohol with students should definitely be done, if at all, only in large groups and never with students who are minors.

TENURE EXPECTATIONS

Tenure expectations vary widely across different teaching schools. All institutions will judge your tenure case on some combination of teaching, research, and service. Because each institution has different expectations and values, I want to provide you with specific questions to ask your department chair or academic dean so you can ensure that you stay on track for tenure. If your institution does not provide you with annual feedback, these suggestions are particularly critical. I encourage you to meet with your department chair, dean, or both, annually to check in on your progress toward tenure. If your institution does not provide you with a senior faculty mentor, it is also wise to find one yourself, who has very recently gone through the tenure process and received a promotion and tenure. That person can give you a comprehensive summary of the current tenure process. Although your more seasoned colleagues who received tenure many years ago can be a fountain of invaluable advice, they are not going to be the most informed colleagues about the current tenure process unless they are on the tenure and promotion committee.

TENURE INTERVIEW QUESTIONS:

- How many peer-reviewed publications are expected for tenure?
- Do the specific journals I publish in effect how the tenure and promotion committee evaluates my publication record?
- Do publications that I co-author with a previous mentor count less than independent publications?
- Do studies that I conducted before my appointment at this job count toward tenure, or do you only count studies that I conducted from start to finish at this institution?
- What are the committee's expectations for conference presentations?
- Are presentations of student-led projects at undergraduate or regional conferences weighted similarly as my presentations at more prestigious conferences in my field?
- Are publications of student-led projects in undergraduate journals weighted similarly as my publications in more prestigious journals in my field?
- Does including a student as an author on a conference presentation carry more weight than if I do not have a student author on the presentation?
- Do publications with a student author carry more weight than publications without a student author?
- How do student evaluations of my teaching factor into tenure?
- Am I required to submit teaching evaluations from my colleagues in my tenure dossier, and if so, how many? If not required, may I submit them?
- What is the format of the tenure dossier? Is it all electronic?
- Can I see the tenure dossiers of the most recent successful tenure applicants?

- Is my current service activity on track for receiving tenure and promotion?
- Are there expectations of service to the community, outside of the university, for tenure?
- How is service to the field, such as serving on the editorial board of a journal, counted toward tenure?
- Is it possible to get more funding for my travel to conferences to support my tenure goals?
- Is it possible to get funding for my student(s) to travel to conferences to present our co-authored work?
- How does advising factor into tenure? Is my time spent advising considered service?

Ashleigh's Interview with Glenn R. Sharfman, PhD

oglethorpe.edu/faculty/glennrsharfman

Glenn R. Sharfman, PhD, is Provost and Vice President for Academic Affairs, as well as Professor of History, at Oglethorpe University in Atlanta. Glenn received his BA in history from Miami University, Oxford, Ohio, and his PhD in history from The University of North Carolina, Chapel Hill. Glenn was the Vice President and Dean for Academic Affairs, and Professor of History at Manchester University from 2005–2014, where I met him when I was hired in 2010. Before coming to Manchester, he served as Associate Dean and Director of Graduate Studies, as well as Professor of History, at Hiram College from 2001-2005. Glenn's family, which includes his wife Susie and their triplets, named Alex, Hannah, and Andy, are an intriguing group of runners, students, political activists, and active community members. I interviewed Glenn because I knew some of his opinions differed from my own. Glenn was an academic dean during my first tenure-track appointment, and I knew he could articulate an administrator's perspective on the value of various aspects of the liberal arts education.

In case you don't know, the dean's perspective is particularly unique and informative for this chapter for a variety of reasons. First, the dean is involved in granting tenure to faculty across disciplines at an institution. Second, they read all the student teaching evaluations each semester, the faculty's annual reports each year, the tenure dossiers each tenure cycle, and the applications for prospective faculty who are interviewing for open faculty positions. Third, academic deans are responsible for the direction of the academic program, so they are in tune with what faculty can do to move the program forward and conversely, how they can get in the way of growth. Therefore, Glenn's comments on how to succeed in your first full-time teaching appointment and how to get tenure come from many, many years of nurturing faculty growth and an eye on behind-the-scenes administrative issues that faculty are typically unaware of.

GLENN'S TAKE ON THE CANDY BOWL

First of all, it's worth pointing out that Glenn hired me into my first tenure-track job, so it goes without saying that he knows me well and watched me develop professionally. This includes my struggle to create a professional identity in terms of how I related to students. Glenn and I both encountered the liberal arts atmosphere for the first time as faculty members, so he understood my culture shock and confused reaction to candy-bowl professors. However, Glenn rightly points out that there is a middle ground within the dichotomy between the heavily personally involved candy-bowl professor and the standoff research professor.

On one hand, Glenn agrees that as dean, he doesn't want professors who are trying to be their students' friends or parents. Glenn supports this position by saying that he and his wife Susie were not looking for someone to be parents to their three kids when they sent them off to college, although he also acknowledges that plenty of students have not had stable family lives and may look to their professors for parental figures. He agrees that it is important for faculty members to know their boundaries with students, although it's not uncommon to befriend or continue to mentor them later in life when they are no longer your students. Glenn dislikes the need to be liked for the sake of being liked, which is reflected in my candy-bowl analogy, both inside and outside the classroom. However, what Glenn sees as the middle ground between being a candy-bowl professor and being aloof is being *interested* in the students.

How a professor demonstrates that they are interested in the students may significantly vary from one institution to the next. At Oglethorpe, where Glenn is currently Provost, approximately 42% of the students are first generation college students and many of them come from inner city Atlanta. That means many of their students come to college without having parents who can provide advice on how to navigate college. Some students do not have a high school education that properly prepared them for college-level course work. Glenn describes the student body in this way because, although at some institutions you can take a sink-or-swim attitude with students and expect them to be fine (because they have a support system in place through either parents who attended college, or private high schools that were considered college prep), that's simply not true at other institutions. It might be too much of a leap from high school to college to expect students to succeed without additional support. This translates into many different opportunities for faculty to demonstrate that they know their students and express interest in them. It could mean taking some extra time at the beginning of the semester to describe the differences between college and high school. It could mean describing how to act in class and how to study so you don't lose them a few weeks into the semester. It means meeting the student where they are in order to put scaffolding in place to add value to their experience, without changing or lowering your expectations. It might even mean attending sporting events to show the students that you care and are interested in who they are outside of the classroom. It does not have to mean being a candy-bowl professor.

TEACHING

As someone who has been involved in hiring, developing, and supporting hundreds of faculty over the years, Glenn has been reading teaching evaluations for thirty years. He sees all the drawbacks of using standardized student evaluations of teaching, including the distinction between being popular versus good, and the unfair treatment of men versus women. I was surprised in our interview that Glenn remembered a disparaging comment a student made on my teaching evaluations five years ago because it turns out that he really does read (and remember!) the evaluations of his faculty. In the evaluation, a student said something horribly cruel like "If Dr. Maxcey wants us to respect her, she shouldn't wear sweatpants to class." I was obviously mortified, but for the record, they were Athleta pants with pockets.

This was an evaluation from a fall semester class, and over winter break, I was throwing my youngest son Henry's first birthday party and Glenn was invited. I'd read the evaluation earlier that day and was hoping Glenn had not seen it. Of course, Glenn mentioned it at the party and said something kind like, "I know you had a baby this last year and I don't care what you wear to class." Glenn's kindness and support of me in response to this comment illustrates that when deans read your evaluations, they are reasonable and recognize when something a student says is not useful or is even cruel. In light of this, his recommendation is that you process your evaluations with both your peers and your department chair.

Glenn pointed out another type of comment on student evaluations of teaching (which also happened to me): complaints about faculty members bringing their kids to class. I've had to bring my kids to class when they had a fever and couldn't go to school, or elementary school is cancelled or there's a fog delay (which happens all the time in northern Indiana). Glenn pointed out that having class with a possible minimal distraction from the kids is preferable to cancelling class (which is something that does actually interfere with the students' learning). All this is to say that students' evaluations of teaching can have little to do with your teaching, and administrators know better than to worry about some of the minor comments.

Glenn is also fully aware that teaching evaluations can be a popularity contest. Some students think professors are cool if they are hipsters, or swear a lot in class, or are very different from the students' high school teachers. These kinds of traits often elicit high evaluations from students. Glenn said that what teaching evaluations are useful for is figuring out who the terrible teachers are. He said that when 60/100 students are saying the teacher is terrible, it is probably true. The problem is that Glenn says he already knows if someone is a terrible teacher and the evaluations don't help the professors improve. But Glenn adds that however flawed teaching evaluations are, they do indicate which faculty are struggling in class. His point is that they do not help the faculty member turn it around. His analogy for this is that when a professor gives feedback like "be more clear" or "awkward" on a student's paper, although it might make the teacher feel good, it hasn't actually given the student useful advice on how to improve. I completely relate to this sentiment; I once provided a colleague with feedback on a grant and circled a sentence as "unclear." The colleague got irritated with me and requested an example of a sentence that would make more sense because it was already perfectly clear to my colleague. Likewise, Glenn says that when students say a professor is too hard or should be more understanding, that doesn't help the professor know what to do differently in the future, even if it undoubtedly makes the student feel better.

Glenn is interested in data that shows that students are learning something, such as measures like pre- and post-tests in a class, or scores on a major field test (i.e., a comprehensive exam on a field, like psychology). Glenn echoed my sentiments that students don't realize what you did for them until years later; at one institution where Glenn worked, the dean solicited letters from graduates for a faculty member going up for tenure. He also suggests writing your own evaluation questions that are tailored to the class, such as asking for feedback on a particular book you assigned that semester. The questions on student evaluations of teaching are often universal to the institution and end up missing the opportunity for specific feedback.

Given the problems with student evaluations of teaching, Glenn encourages faculty to evaluate one another, but he said it's hard to get them to be honest about one another's teaching. This reminded me that when a peer evaluated my teaching, I asked her specific questions beforehand. Because I admitted I was struggling with something in advance, she seemed comfortable enough to brainstorm ways I could improve. By asking her to comment on something specific in her evaluation, it opened her up to give me the honest feedback that I really needed. Glenn says this example is why he prefers peer teaching evaluations as developmental rather than summative.

Despite some difficulty in giving honest feedback to colleagues, Glenn emphasized that attending peers' classes offers a great learning opportunity. For example, Glenn is interested in moving away from the typical lecture style format, in which professors stand front, spewing information at students who are expected to sit quietly and listen. One time when Glenn was evaluating a professor, he was really impressed with how she moved from table to table, interacting with her students. He also decided that it looked exhausting and despite his interest in her style, he wasn't going to be good at mimicking it. This example illustrates that much of good teaching is trying to accentuate what you are good at and do less of the things you are not good at. Ultimately, it's akin to my reaction to the candy-bowl professor; figure out your style and ask for help along the way.

SERVICE

At many SLAC institutions, service is traditionally broken down into service to the department, service to the institution, and possibly service to the community. Many of these service requirements involve going to committees you're appointed, which sometimes require quite a lot of work (e.g., assessing the core curriculum) and other times require mostly listening and reporting back to another group (e.g., update your department on what happened at the monthly technology committee meeting). Although these service appointments can be necessary to running the institution, it does often boil down to checking off a box. Glenn likens these service requirements to a student who merely attends class without studying and gets a C. In other words, this isn't the kind of service Glenn finds impressive when it's time to go up for tenure.

The kind of service that Glenn is most interested in seeing his faculty demonstrate is showing support in various ways that are neither assigned (e.g., committee work) nor expected (e.g., attending a talk when you invite the guest speaker). For example, Glenn is interested in showing up to support talks for members outside your department. If a historian is giving a talk, he really wants to see the biology faculty in attendance, in addition to the history department. Another example is showing up to events where students or other faculty members are performing. Although Glenn certainly doesn't keep attendance at these events, and has reasonable expectations based on what else is going on in his faculty members' lives, Glenn's administrative role has him focused on these gestures that set his liberal arts institution apart from large state schools. He explained that since state schools are often lower in price than liberal arts schools, and may be better able to advertise the fame of their faculty's research accomplishments, he needs to make a case for why students should come to his liberal arts school. He says that what state schools can't provide for students is the number of interesting faculty members that students have the opportunity to get to know as individuals and who are reciprocally interested in getting to know the students. Reflecting on this pivotal difference between state schools and liberal arts institutions, Glenn pointed out that his three children all chose to attend small liberal arts schools (and not at his urging; they applied to large state schools as well). He thinks that their ultimate decision to select smaller schools was that they saw a true, measurable benefit in the opportunity to engage in real relationships with their teachers on an individual level.

RESEARCH EXPECTATIONS

I think the most surprising aspect of my conversation with Glenn was that he gave me a definition of research. In retrospect, it made sense that an administrator from a SLAC would be inclined to define research and his definition highlighted so much about the difference in research at a liberal arts institution versus a research institution, and explains why research activity is often referred to as *scholarly activity* at these schools. Glenn defined research as work that reaches others and helps them in their work. He added that doing research is rewarding and creating knowledge is what we as faculty members were trained to do in graduate school. Indeed, contributing to knowledge is something we all benefit from because it informs our teaching, our research, and our well-being. Glenn emphasizes that teaching and research are not mutually exclusive.

This more wholesome approach to what constitutes research illuminates Glenn's larger point, which is that there are levels of research that we often do not think about, but count toward tenure, and more importantly, those parts matter at a liberal arts institution. At liberal arts schools, many people want to conduct research because good research informs teaching rather than because they received graduate training in the topic. Glenn emphasized that such research often serves as second level, or middle ground between no research at all and research programs that are actively publishing. One popular example of this second level of research is the scholarship of teaching and learning, which is perhaps not as respected among some top tier scientists but can be very useful at a liberal arts school. In Glenn's time as a dean, he has seen some faculty conduct this type of research based on their experience in the classroom, and publish it in teaching and learning journals, or present it at regional teaching conferences.

Glenn considers such projects to be an excellent research endeavor because it's effective at fulfilling the goals of research, according to his definition, and it has the added bonus of being something that the institution can use as a selling point to prospective students and their parents. This example of faculty interest in ongoing professional development is value added that a liberal arts school can demonstrate to set themselves apart from competitive state schools.

If given the choice to create a department of four faculty, for example a psychology department, Glenn said he wouldn't want all four department members to be professors who offer students research experience in a typical laboratory setup like mine. Glenn said he'd love to have one faculty member who does that, but the reality is that because not all students are going to be best served by that opportunity, it wouldn't make sense to have all the faculty offering the same type of experience. For example, not all students would be capable of such an experience because to be productive in a lab setting, they need to be above average students; similarly, students who do not have a natural curiosity about the topic and research process won't be interested in working in a lab. The types of opportunities that I offer students helps prepare them for graduate school, but not all students want to go to graduate school, so other types of experiential learning opportunities are necessary. If all four members of this ideal department were offering similar laboratory research opportunities, the department would fail to serve the other students who don't want to be in, or don't belong in a lab. Glenn points out that a liberal arts school has a commitment to cater to all the students and reach across the vast differentiation of the student body. In order to reach all students in the class and help them move forward through the program, graduate, and find employment, the school needs to offer diverse of experiences. In addition, Glenn would need all four hypothetical faculty members to be really good at teaching Introduction to Psychology because introductory classes are so important for teaching students what a field really is. This rings true for me because my favorite part of teaching Introduction to Psychology is to debunk the idea that psychology is just therapy.

In this ideal department, Glenn would want faculty who could offer experiences that look very different from each other; they might differ in terms of whether they publish research, attend conferences, travel to collect data, conduct focus groups, interview folks on or off campus, seek grant funding, or get involved in community outreach. This ultimately means that to reach his goal of having faculty who can serve a diverse student body, Glenn is unlikely to say that to get tenure, faculty need a certain number of publications or conference presentations or grant applications. He's looking at the much bigger picture of serving students' diverse needs. For me, this ties back to the candy-bowl professor because Glenn is saying there is no one size fits all. This explains why research is called scholarly activity at liberal arts schools and why looking at the current faculty at an institution isn't a good way to determine your fitness in a program (something mentioned in Chapter 6 on getting a job at a liberal arts school).

Despite supporting a range of scholarly activity, Glenn echoed my advice about staying marketable. Glenn said that what sometimes happens at teaching institutions is that people get hired, fail to conduct any scholarly activity, become undesirable, and grow unhappy. If these unhappy professors can't get a job anywhere else because they have not stayed marketable, it really hurts the students and the institution. Glenn would like all his faculty members to be marketable, which probably means publishing and being able to be promoted. Glenn thinks that some turnover at an institution, having people coming and going, is actually good for the institution.

HOW MUCH IS ENOUGH?

Faculty members who know that they will be judged on teaching, research, and service accomplishments when it comes time for tenure, ask the very common question, "How much activity in each category is enough?" Glenn's unique perspective on service and research are complimented by his unique approach of having different expectations for different faculty members. He admits that all deans don't take this perspective, but in Glenn's experience, he can best maximize each person's potential by recognizing that it doesn't make sense to ask everyone to be good at everything. Beyond teaching (because you do have to be an excellent teacher at a SLAC) some people spend much more time and energy on research and some people spend significant time and energy on service. Glenn pointed out that he wouldn't ask someone to run a research lab, serve on many different committees, and go to all the baseball games. He knows that's just unreasonable. Glenn takes the same liberal arts approach with faculty that he takes with students—that one size doesn't fit all,. He recognizes that you just can't ask for more time than a person can give. Because the primary job at a liberal arts school is teaching classes, it just seems silly to say faculty have to be perfect in all three categories (teaching, research, and service). Therefore Glenn tries to sit down with his faculty and set reasonable goals while figuring out how he can best support the faculty and their goals. This allows him to accentuate the strengths of his faculty and steer them away from areas where they'd be less productive or even destructive.

Glenn described how research across different disciplines can also vary drastically in terms of what is possible to do while carrying the typical liberal arts teaching load. Specifically, some people have less time to work on projects outside of class than others and some projects take more time than others. Glenn pointed out that some historians need to do research in a remote location, far from town, for six months. Given that this is not an option while they are teaching, Glenn can't reasonably hold such a scholar to a high standard of research productivity. However, he also doesn't want to stymie anyone's love of publishing with overwhelming service requirements because that would interfere with their research and he doesn't think that exceling at research and exceling at teaching are mutually exclusive. Overall, Glenn knows that having a strong track record in two or more of the categories of teaching, research, and service is possible. He values cultivating his faculty's diverse strengths, considering it very similar to trying to add value to a wide range of abilities in the typical college classroom. Glenn also realizes that faculty members go through periods where research becomes more difficult and better administrators will recognize and understand this. For example, research slows down when faculty have small children, when they are caring for aging parents, or other periods when they are more fettered.

UNIQUE ISSUES WHILE GETTING TENURE AT A SLAC

When I asked Glenn what someone should NOT do at a liberal arts institution, he said you should never make fun of your students for not knowing the basic conventions of academia or even how to study. He said that it's normal to notice differences between your students and your generation. Some things can be annoying (e.g., the seemingly never-ending battle with distracting technology), but you have to understand that your students are 18-year-olds. Glenn acknowledges that when he was 18 he was a decent student, but like most others at 18, he was also immature. At that age, students are not fully formed and they don't want, and don't deserve, their professors making fun of them. Be very careful as a young faculty member to avoid speaking negatively about your students.

Glenn also advised you "seek first to understand" and not jump to conclusions about what goes on behind the scenes at a liberal arts school. Almost all faculty members who are hired at teaching schools are coming from graduate programs at large schools, so very few of comprehend what it takes to run a small school. Even if a faculty member went to a small liberal arts school as an undergrad, they didn't really know what was going on behind the scenes of that school, so they do not understand what makes such places unique from an administrator's perspective. I really struggled with this when we were creating a cognitive neuroscience major in the psychology department because I felt confused and irritated that the entire faculty of the school could vote on the major. This meant that I had to justify to the entire faculty from the entire university what in my view, was best for our psychology students. I did not understand why the faculty in the psychology department weren't just trusted to know what's best for their psychology majors. When I reminded Glenn of this example, he explained that most of what happens at teaching schools is part of faculty governance, so it's unwise to jump to conclusions about what goes on in other departments or the motivation of other faculty members. Glenn suggested that you avoid burying yourself in your department if you want to be successful because in instances like the one I mentioned above, it's useful to have strong relationships across campus. In other words, you should transcend your niche, for a variety of reasons. In part, you should be visible across campus because your tenure committee will include people outside your department. Beyond securing tenure, in Glenn's experience, almost everyone thinks that their program is the best and their department works the hardest, but there are so many things that happen behind the scenes of an institution that it's much better not to get caught up in drawing those types of conclusions. Rather, be supportive and respectful of your colleagues in other departments and don't compare yourself or your program to others.

Glenn also argued strongly against the myth that you should keep your mouth shut until you get tenure. He believes that there is no reason to delay expressing yourself until after tenure, so speak up in faculty meetings, answer questions, and ask questions of other programs. In fact, doing so demonstrates leadership and leadership potential. Glenn notices the faculty members who he thinks will be great leaders someday, in part due to their behavior in forums like faculty meetings. In Glenn's experience, tenure doesn't change people. People who are productive before tenure are productive after and people who were lazy before tenure are lazy after tenure. This means that if you do not speak up before tenure, the concern is that you will be equally timid after tenure, too. It is therefore a mistake to hide who you really are because you're afraid of the tenure process. It might actually *hurt* your bid for tenure.

Glenn suggested that you get to know your administrators. Glenn tries to have lunch with faculty periodically. He had one faculty member accept the invitation but ask not to have it in the cafeteria because the teacher did not want to be seen with Glenn. This sounds like an unfortunate flashback to a high school homecoming dance experience for most of us, and highlights that often there is a division between faculty and administrators, but you should try to transcend it. Glenn suggested that you should not think of administrators and faculty as *us* and *them*. He said that most administrators are good people who want the best for their institution. But that doesn't mean they will see eye-to-eye with all faculty. Glenn respects faculty who openly disagree but wants them to understand that he needs to do what he thinks is best for everyone.

Glenn said that, for example, good faculty members get to know their department's enrollment person and find out how they are advertising the department. This can allow for course correction if necessary. For example, if they are advertising your animal lab to all prospective students, but in reality, you can only allow a few select students to work there each year, then it's false advertising to indicate that many incoming students have that opportunity. Get to know your Vice President of Communications and send them information for press releases when you publish a paper, or give them the contact information of a student who they could interview for the university's website. It's helpful for both the institution and you to get to know people throughout different levels of the administration. This is such timely advice because Geoff and I were just talking about how we know so much about getting into graduate school, but we actually know comparatively little about getting into undergrad. Inspired by Glenn's advice, I'm going to make an appointment to get to know someone in admissions at my institution.

We discussed other aspects of success, such as advising. Although we don't talk about it much, advising can take a huge amount of time and can be really important. Models for advising differ across institutions, but knowing students have more on their mind than just their courses can be a useful, whole-hearted approach to interacting with your them. Because faculty's involvement in advising varies so greatly from one institution to another, I've chosen not to spend time discussing it in this chapter, but be sure to inquire about the role of advising in your service to the institution, both when interviewing for a faculty position, and in terms of your tenure application.

Glenn also mentioned that he often asked job candidates how they would define success for their major or program, an issue that certainly arises during the tenure process. Glenn said a good answer to this question from an English department faculty candidate would be that a successful program equips students to be in the working world. Glenn said that a bad answer to this question from a History faculty candidate would be that getting students into a good graduate program would be success. Glenn said that an applicant once told him this and he wanted to respond by saying that the faculty position for which this applicant was interviewing had over 300 applications, meaning it's irresponsible to encourage all students to aim for graduate school in history because there are not going to be jobs for all of them. The job market for people who get their PhD in the humanities is intensely competitive (hence the over 300 applicants for the history faculty position). Based on his experience watching faculty applications roll into multiple institutions, Glenn said the reality is that in some areas of the humanities, if PhD students are lucky, they will get a job at a community college. Glenn said that the problem with this candidate's response was the desire to create mini-me's. He acknowledged that he used to harbor the desire to create little Glenn's—other students who he could groom to follow in his footsteps and become history majors. As he's matured over his career, Glenn asserts that now his goal is for the students to become gainfully employed. At liberal arts institutions, the mission is focused on educating and supporting the whole person, but the problem is that's not what most faculty applicants believe when they are interviewing. In other words, yes, you are being hired to teach psychology courses but you're not being hired just to teach psychology. It's more than just teaching and you have to be cautious not to just try to replicate yourself in your students. Glenn described the current state of the job market as one in which most folks are in jobs or interviewing for jobs that have nothing to do with their major. This incredible disconnect between people from their major later in life is consistent with my point in Chapter 2 about double majoring. This also means that the job market is more diverse than the list of majors at your institution, so a faculty member's role is definitely not just to teach their list of classes. Glenn urges faculty members to understand that students are more than just majors, there's more than just graduate programs, and you're going to be their mentor, not their friend. While you don't need to be a candy-bowl professor and you don't have to have them over to your house, you do want to understand that students are not all the same and you need to offer them a diversity of futures.

RIGOR AND VALUE ADDED

Many of us in education love talking about rigor. How to be rigorous in your classes is an issue you need to contemplate as you practice and refine your teaching craft. A huge problem is that knowing how to be rigorous is such an important part of faculty development, yet we are not taught about rigor. When Glenn talked about showing an interest in your students and meeting them where they are, he said you want to do that without lowering your standards. But how do you do that?

Glenn clarified that grading hard is not the same thing as rigor, nor is assigning papers that you know are too complex for your students. To describe rigor, Glenn started by saying that most classes have some bar that everyone needs to meet. Being an accredited institution means to pass a class, students must demonstrate that they have met some standard learning outcomes. However, Glenn believes that rigor involves layering student-specific achievements on top of that bar. He talked about rigor in terms of finding where individual students are in their ability and trying to improve them by the end of the semester.

Glen said that for a student near the bottom of the class, rigor could be trying to teach them that it's important to talk to the professor during office hours. The goal is to teach students that it's ok to ask questions and Glenn takes responsibility for finding ways to make talking to him less foreboding for students. Glenn finds that strong students unapologetically seek out help, but he hopes that even just one intervention with a below average student might show them that he cares and have positive downstream effects for the student. Even if the student still gets a low grade in the class, if they were pushed out of their comfort zone and learned to be more comfortable asking questions, that's value added. Maybe this student did not even meet the basic standards, but to Glenn, their experience in his class was still rigorous.

For stronger students, rigor might look like giving a very strong student a B on a paper because it is not up to their standard, even though it is technically A material. Glenn told me a story of doing just that with a student who then angrily confronted him because she had never gotten a B. Glenn told her that it's true that the paper was not a B paper for the rest of the class and that if she took him to court arguing that it was objectively an A paper, she'd probably win. Glenn told her it was a B paper *for her* because he knew that she could do much better. Her next paper was twice as good as her other work and he used this story in her law school recommendation letters to describe her perseverance and competitive nature. It's not surprising that her update is that she got in to law school. Again for this student, rigor was something Glenn implemented with her that was outside the basic expectations of the class.

Getting to know students so you can implement rigor and supply them with value added is a luxury that can happen more often at liberal arts schools than state schools. Glenn pointed out that if you are at a larger or more research-focused institution, often each faculty member teaches only a couple of classes with a number of sections. This means you may know a student for one or two classes. You might meet them when they are at the start of college and never see them again, or you might meet them when they are fully formed as a junior or a senior. An example of *value added* is when students are seen by the same professor throughout three or more classes, over three or more years, so they get to know a student well. The professor can see their growth, they can vouch for it, and they can treat them differently (in a positive way). This is when the opportunity for differentiation among students occurs, which enables rigor and value added to emerge.

ADMINISTRATOR'S ISSUES

Geoff and I would love to have another chapter in this book about becoming an administrator, but we will have to save that for another book. In my conversation with Glenn, he mentioned some aspects of being an administrator that I think can be useful for those of us who don't know much about administration, so here are a few random administrator issues that I wanted to share with you.

First, Glenn has always philosophically raged against grades for as long as I've known him. He spoke about how we make so much of grades but he doesn't think they are representative of the most important aspects of student development. Glenn said that he loves his C students and that when a student earns a C, it doesn't mean he is making an assessment of their worth. Unfortunately, grades are tied to so much, including scholarships, so rather than a focus on actual learning, the pressure to get good grades, is really intense. I think it's great to know that administrators like Glenn are even thinking about these issues.

Second, Glenn feels passionate about living what he says he wants from his faculty members. He's always practiced what he preached, and I think that's made him a great administrator. For example, he shows up at basketball games when he wants faculty to do it. Glenn believes that some administrators go into administration because they were subpar teachers and they speak to their faculty in data, but that's not what Glenn thinks faculty need. Glenn believes that faculty need a cheerleader and someone similar to a coach—someone who has a playbook and a goal for the team and knows who to kick in the butt and who to pat on the back. This coach analogy rings true for me when I need encouragement and support for my career vision, but I also want someone to have an eye on the larger mission of the institution that I serve and help my actions align with the bigger picture of the institutional goals. This is part of administrators' role and why it's so useful to get to know them.

Third, it's obvious that Glenn really wants to leave a place better than he found it. One of the things he has some control over that has long term impacts is hiring decisions. At liberal arts schools, hiring decisions are not as cut and dry as they are at research institutions where the main criteria are number of publications and amount of grant funding. In order to get the right people to move the place forward, you have to really know your institution and be in sync with its mission when you hire faculty. I remember quite clearly most conversations I've had with Glenn when he was my dean at Manchester. Some of them were tough conversations to have, but I always felt like he really got me, he took the time to understand my vision and my career goals, and although he sometimes had to help me course correct (I was in my twenties, ok people?!), I always knew he had my back. Through daily actions like these, there is no doubt that he left the institution better than he found it. This summarizes everything you should want from your administrator.

CHAPTER 9

How to Succeed in Your New Faculty Job and Get Tenure at a Research Institution

by Geoff

Welcome to the show! For many of us, a tenure-track job at a major research university was the goal of the last decade or more of our lives. If you now have one of these jobs, you have cleared a very large hurdle.

There is more good news. For the most part, if you keep doing what got you there, you will be fine. Here, I will discuss some new stuff we all had to learn. Most people are not used to running a large classroom like you will need to do as a faculty member. Advising multiple trainees at the same time will likely seem new compared to the small-scale advising you did of undergrads or junior graduate students in the past. However, the new demands are not really that difficult to learn, as they are a matter of scaling up what you have been doing. If you have been following our suggestions, tips, and advice, then you will likely be fine.

The most dangerous thing for your career success in this phase is your own fear of the process. Just continue to publish, submit grants, and everything is going to work out. Some of us nearly lost our minds pre-tenure, but you do not need to be so neurotic. Stay focused on one paper and one grant at a time. Proceed with confidence, as a former mentor used to advise to me.

Essentially, all major research institutions make tenure decisions based on your success in three areas; research, teaching, and service. One of these is more important than the others. Can you guess which?

Research is the most critical. The chair in your new department may be frank with you about this priority setting. However, it is possible that they will not be this blunt because they really hope that you will be excellent at everything. Let's look at the materials you submit when you go up for tenure at a major research institution to make sure you understand why research is the most important component.

The set materials that are submitted when you go up for tenure is sometimes called your *packet* (or *package* at some institutions). It has your CV, copies of your published papers, your teaching evaluations, and external letters from tenured faculty in your field. Typically, you recommend about half of the letter writers and the department comes up with names for the other half.

The external letter writers are asked to comment on the impact and quality of your research. This is why research is so important. Research productivity and impact in the field is what your external letter writers are asked to assess. These letters often discuss the impact of specific papers. Letter writers are also often asked to comment on whether you would get tenure at their school. Some letter writers will also comment on the total number of publications in the pre-tenure period, or the average number per year.

Why would someone comment on the number of publications per year? This is relevant, and research offers a concrete guide here. Several years ago, a paper in *Psychological Science* showed that people publishing approximately three papers per year during the pre-tenure period get tenure[14]. This was across multiple areas of experimental psychology. This is also consistent with the number that Marvin Chun told me when I was a postdoc. What he learned by observation across his career was that we need about three papers per year to be competitive for good R1 jobs, and to get tenure at R1 institutions. If those papers are in high-impact journals, all the better.

Some methods in neuroscience are much more time intensive and result in significantly lower rates of publication. One example is awake behaving animal neuroscience. In general, one independent paper per year is considered a good rate for junior faculty with a small animal lab. Developmental science often involves rather demanding data collection efforts such that publication rates are also lower than in many other areas of cognitive psychology that study healthy young adults. Finally, cognitive modeling typically results in a slower publication rate, due at least in part, to the slow review and publication processes for this work at the most relevant journals.

After all the materials are together, the tenure committee reviews your letters, papers, grants, teaching evaluations, and your record of service. Then they make a recommendation to the department. Most of the time these recommendations are affirmed in a department-wide vote that follows a lengthy discussion.

In sum, due to the nature of the external letters, the assessment of your pre-tenure period is focused on your research. You cannot just blow off teaching or service in many of the best departments, but part of the difficulty of time management is to understand how you can teach in a time-efficient manner so that you can maximize your time spent writing papers, grants, and in the laboratory.

RESEARCH ABOVE ALL

Because of the importance of your research for getting tenure, the first thing you will want to do is to get your equipment set up and get your Institutional Review Board or Animal Care and Use Committee proposal done, so that you can start collecting data. This is necessary because you need pilot data from your lab to put in your grants. In addition, you need something for your new grad students, postdocs, and research assistants to do. Note that this setup time is typically uneven, in that you will work on it for a day or two until you hit a roadblock and figure out what the next thing is that you need to order. This makes equipment setup well paced to begin immediately, and then interleave with the writing that you need to do, as I discuss next.

If you followed the advice from the postdoc and graduate student chapters, you have already started writing a grant and really just need to add some more pilot data from your own lab to it. If you haven't worked ahead like this, do not panic. The grants that fare the best during the review process are those that describe a line of research that is uniquely your own and that you have demonstrated your ability to conduct. This means that you could start writing this grant even before your equipment is completely set up. Start writing a grant based on your primary theme of research throughout graduate school and your postdoc. If you want to change directions, then what you need to do is run experiments and publish a few papers before you are really ready to write a grant on that new direction. A demonstration of feasibility can sometimes be done with large amounts of preliminary data in a grant (i.e., several different experiments and analyses). The surest way of appeasing reviewers' concerns about feasibility is publications. This means that you should focus on running experiments and writing up papers.

Jon Kaas is a wise senior faculty member at Vanderbilt, and a member of the National Academy of Sciences. His saying is that every new faculty member should have a list of five things they could do tomorrow. These are five things that are likely to yield publishable results, regardless of which way they turn out, and that you could run yourself without difficulty tomorrow. These are often follow-ups on things you started in grad school or as a postdoc. You might also have a number of ideas that you've accumulated during your training, and hopefully, you have these written down somewhere. If you experience freezing like a deer in headlights because you feel overwhelmed, then you should dig out those notebooks you began in graduate school that you've been writing your ideas in and do those five straightforward things.

LAY OUT YOUR PRIORITIES IN A GRANT PROPOSAL

As we describe further in Chapter 10, your first grant will likely be on a topic that you already have published on extensively as a graduate student or postdoc. This is necessary because you want to be an expert on the topic and methods described in your grant. You might be thinking, "I do not even have my lab equipment set up; how can I write a grant?" Actually, this is the perfect time to determine what the first experiments you need to run are once that equipment arrives and is operational. By writing the grant now, you can determine which experiments require pilot data for your grant. Then, you can collect and analyze those data and drop the preliminary results right into the grant's figures.

At this early stage, your grant can serve as something of a blueprint to keep your research on track. It is easy to become distracted by a recent publication that seems impossible or some funky new effect. You might think that your new lab should replicate that impossible finding or follow up on that funky effect. This type of project is what good scientists occasionally pursue when they have the capacity, but this is not the cornerstone on which fundable lines of research are made. Instead, you want to have a systematic line of work that you grow and cultivate into a research program that can sustain multiple personnel.

It is easy to get the steps backward. That is, you plan to set up the lab and start running experiments first. Then you can see if you discover anything interesting and write a grant on that finding. This is a strategy that might work if you discover something cool right away, it is published very quickly, and it is theoretically important enough to convince the reviewers of your grant that it is a good funding bet. This strategy is not a good bet in general because you put the creative process of scientific discovery under time pressure that does not recognize publication lags and other things out of your control. You may also face a threshold problem in which you never feel totally convinced that a result is novel and interesting enough for a grant. Finally, reviewers will want your CV to demonstrate that you are one of the world's experts on a topic and on the method you use. So my advice is to go ahead and write a grant on what you are an expert on; the stuff you have been doing as a grad student or postdoc, with some novel twist. Down the road, you can write grants on the cool stuff that you find during your first 5–10 years in your laboratory. If you do it the other way around, you are unlikely to have the lab running long enough to get funding for those cool things that you find during your first five years or so.

Do you really need to get a grant to get tenure? Your chair and department members can provide further guidance on your specific circumstances. There are definitely universities at which the inability to fund your work would be a red flag on a tenure packet. However, I have seen people at major research universities get tenure without a grant. These are generally people who have published a large quantity of papers despite the absence of external funding. Typically, I've seen this situations when the faculty member's lab runs experiments with human subjects from the credit pool (i.e., free research participants). The point is that your institution wants to tenure people who have had a significant impact on the field. Getting a grant can help make that possible. But you will have to get papers out in the literature. Some people have found success dedicating themselves to writing as many papers as possible, instead of writing grants. This brings us to our next, more general topic of time management.

PROTECTING AND MANAGING YOUR TIME

Being an assistant professor is really exciting. It is all shiny and new. You get to be the one asking the students hard questions in committee meetings. You get to influence how things are run in yet other committee meetings, and you get to have meetings with your own students. As you can tell, it is fairly easy to end up with a calendar filled with endless meetings.

At first, I found that the easy thing to do was to spend all my time talking to people and little time writing. Later, I learned that to get my reading and writing done, I had to hide from the other aspects of work. People do this in different ways. They might work from home, have a coffee shop they hide in, or simply close their door and not answer if they hear a knock. One of my senior colleagues liked to jokingly introduce themselves to me each time they saw me, as a comment on how well I protected my writing time. The implication was that it was hard to drop by my office and talk whenever they wanted. However, as Ashleigh mentioned previously, it was necessary that I protected my own writing time. No one will drop by your office and ask if they can help make sure you have a couple hours each day to write, program, or analyze data. You have to do this for yourself.

I would recommend blocking off time on your calendar for writing, preparing figures, and analyzing data, and set this to repeat each week. If I fail to do this for myself, then I get an email from someone requesting a meeting and I give away that time because it appears to be open on my calendar. Getting papers out is not some optional thing that would be great to get to when everything else is done. It is the priority and should be treated as such.

Protecting your writing time is not just good for you, but also for your trainees. Faculty members vary in how quickly they get manuscripts back to their trainees. It is very easy to slip into thinking that a turnaround time of several weeks or even months is good enough. However, this does not reward your trainees for the hard work they did running the experiments, analyzing the data, and getting you the first draft of the paper (no matter how rough that draft is).

When I first became an assistant professor, I used to keep track of my turnaround on manuscripts, just like journals track time since submission. For a while the mean of my turning around time was a few days. This was probably too fast. Some papers require you to read them, think about them, revise them, think some more, and revise again, before sending them back to your co-authors. I have not found it possible to think faster about theoretically complex issues, no matter how hard I try. So some papers and grants might take a few weeks, because you go through multiple iterations by yourself. It is always a good idea to shoot an email to the first author (e.g., your trainee who wrote the first draft) to let them know what is happening. I have found that students and postdocs work hardest when they know that you are going to put their manuscripts at the top of your priority list when they send them. This keeps these manuscripts coming as fast as possible, and this will get you tenure.

Getting a manuscript done quickly can cast a shadow. While you are trying to get papers out as fast as possible to secure tenure, it is tempting to simply fix papers without training your people how to write better themselves. This can result in your trainees shooting you manuscripts that are half baked and not proofread because they know you will just fix it for them.

To avoid the problem of becoming a paper repair technician, I have two suggestions. First, use your word processing program's comment feature heavily when you edit, particularly for the first several papers you work on with a trainee. If you change or rewrite the text, explain why. I have colleagues who take this a step further and sit down at the keyboard with their trainee to rewrite papers together. I find this torturous, but it may be useful for you. I use the comment feature in my word processing software to explain the thinking behind my revisions so that trainees can learn to write better. You could also revise the paper and then meet with students to step them through your logic, but my memory is typically not good enough to remember why I made each of the hundreds of changes in a file.

My second suggestion is to tell the trainee that they should imagine that they are not sending the paper to you, but instead sending it directly to the journal. Usually a trainee is capable of writing a better paper, but they are trying to get it to you quickly. If you let them know how you approach the writing process, that is, you write a draft, set it aside for a couple of days, and then come back to it, with perhaps three or four iterations before you send it to anyone, they will know that it is good that they take their time and go through several drafts.

Finally, it can be useful to have lab meetings about scientific writing. There are several excellent books on the topic, including a few by Robert A. Day[1] that I have used in my lab meetings (along with the classic by Strunk & White)[2] to explicitly lay out the rules of scientific writing. The first time I went through those books I found it transformative. I hope your trainees will too.

WHAT WE DIDN'T TEACH YOU

The example of teaching scientific writing that I discussed above is just one of the things you are asked to do as a faculty member that you probably were not trained for. For most of these kinds of things, we do what our mentors did, but some things, you didn't get to see. How do mentors manage their budgets? What characteristics do they look for in potential RAs, grad students, and postdocs? How did they handle getting research done in the early days, before you came long? How did they handle complaints from undergraduates about grades? There is an infinitely long list of stuff that you were not taught and will be required to figure out. I would like to talk about a couple of these, but I hope that you take an opportunity to talk to your colleagues and mentors about the business of professing. I also encourage you to make friends with your colleagues in the department when you start. Go to lunch with them every week or so and find out how

---

[1] *How to write and publish a scientific paper* by Gastel & Day and *Scientific English: A guide for scientists and other professionals* by Day & Sakaduski

[2] *The elements of style* by Strunk & White

they handle these issues. Frequently we recreate the same solutions to these common problems ourselves, when we could have just asked someone.

You are probably going to run your lab for several months before you have real help. In a worst-case scenario it could even be a couple of years. I was able to recruit graduate students during my first year as an assistant professor and got some good people right away. However, I was still in the lab collecting data whenever possible for about three or four years after I started my faculty position. Some people I have talked to say that they did not do any of this. They got great students right away. Great students and postdocs kept coming. However, I do happen to know that these faculty are in the minority. Most of us were down working with the gear for a couple of years before our students really edged us out. This is necessary. It is also not a bad thing. I loved being in the lab as a student and postdoc, and now it's all mine. One of my amazing colleagues who recently passed away spent her entire career in the lab as much as possible. When not teaching or working on a paper, she was at the bench doing the work with her staff and students. Since research is the exciting part of the job that we fell in love with, then why do we leave it? You don't have to leave it, and in the beginning, you are needed in the lab.

It is possible that your first couple of students won't fall in love with research and won't be productive. For this reason, it is very good advice to hire an RA right away. Email your friends and put out advertisements as soon as you know your start date. You will still need to formally hire them through the university's human resources procedures, but begin this recruiting right away (as I mentioned in Chapter 7). Ideally, you should be able to get tenure based on your work alone, with the help of your RA (or undergraduates) to aid data collection. It almost never comes to this extreme situation in which you do not have grad students or a postdoc in the pre-tenure period. However, thinking that the recruitment of an outstanding graduate student is the answer to all your problems is not wise.

How do you know who will be amazing? I wish I could accurately predict this. People usually base hiring decisions on a candidate's interest in the work you do. It is useful if the candidate has been doing related work, so that they really know what you do and the landscape of the field that you are in. Unfortunately, most metrics like GRE scores and GPA are only weakly predictive of success in graduate school and beyond. Many people prefer to rely on past productivity as the surest sign of future productivity. That is, if your candidate has already been publishing papers, they are likely to keep doing that. This is why people favor making graduate admission offers to people who have worked as full-time RAs prior to graduate school, during which time they obtained authorship on a paper or two. When recruiting postdocs, productive folks are highly sought after because they know the amount of work required to churn out two to four papers per year before they get there.

As I said at the beginning of this chapter, the great news is that most of what you need to know to get tenure you already know. That is, you know how to publish papers. Your publications are the key to getting tenure. Those same publications result in successful trainees and the grants that you need to support your ability to keep publishing. My belief is that the greatest obstacle to getting tenure is our own meltdowns during this stressful, early period. I had to constantly remind myself that this was the same game that I fell in love with during graduate school and my postdoc. I had to remember to have fun with the projects, writing grants, and working with the other people in the lab.

ADVISING TRAINEES AND SETTING EXPECTATIONS

If you are like me, you are planning to train your students similar to the way your graduate and postdoctoral advisors trained you. There is nothing wrong with that. In Chapter 5, we discussed the need to explicitly think about how you want to mentor. Do you want to have weekly meetings with each trainee? Do you want to spend an hour or two in the lab with them each day? Or do you just want status updates via email when they need something or have news, outside of a weekly lab meeting.

My reading of the literature on mentoring styles suggests that there is not a single ideal way to mentor everyone. If there were a best answer, I'd recommend that. Instead, mentoring tends to be an individualistic characteristic; you can pick a mentoring and management style that fits you. My experience is that most mentors meet with each of their trainees weekly in individual meetings and have a group lab meeting every week. Some mentors have additional meetings periodically, such as every couple of months, with groups working on specific projects. Although that is the modal schedule in the field, people's mentoring schedule can change across time.

I started off having weekly meetings with trainees. Then I found that for trainees who did well, these meetings were unnecessary because I would talk with them almost every day when I was down in the lab. For these productive people, the scheduled meetings were a waste of time that kept them from collecting more data and we decided it was more efficient to just not have them. For the people who struggled, these meetings exacerbated their difficulties. The struggling trainees would work just enough to have something to report in the meeting. This included working until they hit some predictable problem. Then, at our meeting five days later they would ask how to get around that problem. Despite my advice to them about problem solving and encouragement to start collecting data and publishing papers, these weekly meetings seemed to set a very low bar because these students would stop at the minimal amount of reportable progress each week. Thus, the weekly meetings seemed to help the weaker students become even weaker. At the same time, the strong students would talk to me multiple times each week outside of such scheduled meetings, so the meetings were unnecessary.

On an emotional level, weekly meetings seemed more painful for me than helpful for the struggling students they were meant to help. Perhaps this is a personality trait that leaves me at a deficit. That is, my unproductive trainees seemed to not to be motivated to show me what they could do. They were instead confident enough in themselves that they were totally comfortable saying that they hadn't gotten anything done that week in the meeting, or that they were busy with classes this semester, so they couldn't plan on doing any research for the next few months. Clearly, this is not the case for most people, but my experience is that the people for whom the weekly meeting could be the most useful, turned out to find them the least useful, and even seemingly harmful.

For the reasons just discussed, I have developed a different schedule for talking to trainees about their projects. I now have initial meetings with trainees in which they discuss what they should do right away. I give them some papers about the issue or phenomenon that the new experiment examines (sometimes this is directly from a grant that they are also given to read). I describe the experiment that I would like to have them run and get their input about whether this is something they are interested in, and whether they have ideas about the experimental design. If they are more interested in some other line of work that we have going, I am fine discussing alternative projects. I really want to find a good fit between a project they are excited about and a project that we are well positioned to run.

After we settle on a project, I tell them to let me know as soon as they have a version programmed. Usually this is something that can be done within a couple of hours to a few days' worth of work, assuming the trainee is coming in with programming experience. The students destined for success get back to me in a couple of hours with a version done and some high-level questions. Sometimes those whom I have to hunt down have completed the task too and are already on to the next step that they figured out themselves. However, sometimes the trainee will have gotten stuck and not sought help. I will talk about these struggling cases in greater detail later in the chapter.

In the example above, I talked about launching a new student on a project alone. I need to note that this is not the norm. Typically, I will pair a new trainee (graduate student or postdoc) with an existing lab member so that they can learn how to use the equipment. This is the norm in the field, from my experience. If the new lab member does not know any computer programming, this pairing is critical. The hope is that people without programming skills will begin by collecting data and then learn (i.e., teach themselves) how to write and use code during this first project, in which they receive scaffolding from the existing lab member and me. Sometimes it works, and the trainee learns the necessary skills during that first project and becomes increasingly independent across time. However, it doesn't always work out this well. Running that first project without carrying the weight of the programming and analysis can result in setting the wrong expectation: that programming is something you have other people do for you. Be explicit that the first project they run with someone is just a learning experience, and that you expect them to set up and analyze the next experiment.

I want to note that graduate students who enter a lab without any programming skills have a much lower probability of success than those who learned programming as undergraduates or RAs. The probabilities are different enough that many potential mentors that I know do not make offers to people who do not already know some computer programming. Be warned, if you take a graduate student or postdoc into your lab without such necessary basic skills, the person will need extra resources, and still, the odds are not in their favor. As a mentor, you should advise your undergraduates and RAs to take programming classes (either at your institution or online).

I tell my undergraduate advisees that if they would like to go to graduate school, the only class that really matters is a programming class. This is hyperbolic of course, but in a practical sense they will forget most of what they learned in their class on western civilization while getting an A, but they will use the programming skills they learn in intro computer science every day.

YOUR TRAINEES HOLD THE KEY TO THEIR SUCCESS AND DEMONSTRATE THE LIMITS OF YOUR INFLUENCE

In the last section, you may have picked up on the tone in which I warn you, the mentor, about trying to tightly control the learning of your trainee. This is because the most important factor in training students are the students themselves. Some people are good at problem solving, work long hours until work is completed, and are internally driven. Other people are used to external motivation and look at graduate school either as a 9-to-5 job or like it's no different from being an undergraduate. I have not figured out a way to help the latter type. Sometimes they can be awakened by a serious meeting in which you explain what is required for the successful completion of a PhD program and state your opinion that they are not putting forth the effort necessary to become an independent research scientist. However, I have only seen such a conversation work in a small percentage of cases.

As a mentor who has limited time to run the lab, teach, serve on committees, raise a family, and so on, you need to be keenly aware of your limitations. If we could make people become different people through encouragement, then all the problems of clinical psychology would be solved already. Instead, you should reward people for good behavior and make suggestions when you see them mismanaging their time. But another critical step in dealing with struggling trainees is to document the situation when it looks like it is turning in the wrong direction. When you give them a 2-hour task and it takes 7 days, start a file with notes documenting these struggles.

Your notes are usually vital for convincing yourself about the situation. Once I gave a new grad student the task of installing a piece of software so that they could open a file and look something up in that file. This was a 10-minute task, not the 2-hour version described above, because they had not completed the 2-hour version. The next week, they said that they couldn't get the software installed. I sat down with them and clicked through the dialogue boxes to install it for them. Then, I gave them another 15-minute task. Ultimately, these experiences continued, and the student was never able to fix problems or meet the seemingly routine challenges they faced. This is the kind of thing that goes in the files you keep on such trainees.

Keeping files on all your trainees is a good idea. Your trainees may be very strong and accomplish exceptional things, and you want to keep notes on the exceptional things so you can write them glowing letters of recommendation. If you rely on your imperfect memory to accurately catalog the day-by-day struggles (and successes) of your trainees, then you will have the vague impression that things are not going well. However, such impressions can be biased due to a sunny personality, or even your own competing demands that make it hard to remember what happened in their first year.

The other reason for taking notes on your trainees is so that you can have a conversation with the struggling students in which you give feedback that things are not going well, and you will need to be specific (i.e., the serious meeting mentioned above). Having notes is necessary because sometimes the student is not aware of their own struggles and you need to concretely remind them of a series of events. Worse yet, some students may deny their failures and you need to be able to provide evidence of them. After that meeting, if things continue to go poorly, then you should talk to the director of graduate studies or whoever has an analogous position in your department. You will probably be told the options, which are putting the trainee on probation, or encouraging them to seek out another lab to work in. Do one of these things; do not just put it off, thinking that the person is having personal problems that will resolve themselves and the issues will go away.

Most of us who have struggled at work also had personal problems at the time. The cause-effect arrow is hard to determine, and the arrow is probably bidirectional. But I do not believe you are doing trainee a favor by avoiding a discussion of their lack of productivity. What I usually find is that if nonproductive trainees are allowed to slide, in a couple of years the story changes, and the student tells you that they have invested so much time in the program that they can't fail now. Indeed, needing to put a third-year or fourth-year graduate student on probation typically feels really bad. It is better to figure out that a graduate student is not cut out for a PhD program early on. This is better for them as well. This way they can leave with the terminal master's and avoid spending years toiling in frustration just to fail their qualifying exams or dissertation defense. When people do leave with a PhD but are not cut out for the independent problem solving expected of someone with that degree, things still do not go well. Jobs in the industry typically expect PhDs to be able to fix issues or lead their own teams of researchers. If they are struggling to get basic requirements done under your supervision in the relatively friendly and protective environment of graduate school, then it is unlikely that they will succeed as PhDs in academia or industry.

As I briefly mentioned above, your notes about student progress will be very useful when you approach your director of graduate studies regarding the struggling student.

Telling your colleague that this student seems to have trouble is a vague and fairly useless statement. However, if you have concrete notes, then the department will know that this student is having problems with routine tasks, and this is not a consequence of fit between mentor and student. Often, struggling students try to flee to another lab but when the problems are about a basic inability to problem solve, or an inability to complete straightforward tasks, then your colleagues in the department need to know that.

My experience is that refugees that flee their initial lab have a low success rate in the next lab. Before you have tenure, <u>*do not take on such a student*</u>. Struggling students seem attracted to the laboratories of assistant professors. Such students are probably looking for someone younger and friendly that they can relate to. But regardless of the struggling student's motivation, do not agree to mentor students who are having problems in another lab before you have tenure. A problematic student often makes far more work for you than having no student at all. A talented undergraduate would be better to mentor than a graduate student in meltdown. Allow the more senior members of your department to handle these students if they feel like they can; you should not try to rescue such people while you are also trying to swim to the safety of the shore of tenure.

One of the first students in my laboratory was someone who came from another lab. They were ultimately great and now have their own faculty position. Things did not go so well with the next four. This 20% success rate is similar or better than that experienced by other people I know. When you take on struggling students from other labs you have less space for new people who are genuinely interested in working with you. Unhappy people also make it more difficult to recruit new people. My recommendation is to wait until after you have tenure to see if you can beat the odds the rest of us have experienced with people switching into our labs. Before that, recruit your own trainees and do not gamble with struggling students.

This discussion probably sounds like more detail about how to be a tough mentor than you want to hear during the exciting period of getting your own lab up and running. It is necessary, though. About half of all graduate students do not work out, according to numbers provided by our associations and societies. You can search the internet if you think I am too pessimistic. This means that you need to know how to handle these situations. Unfortunately, you only have your own experience: you were successful and your career went well (although you had your own struggles too). The good news is that you will have students like yourself too. They will do amazing things and solve problems that you were not sure you could have solved. These people fill you with hope and excitement. These people also carry you on their shoulders toward tenure. Enjoy the ride.

COLLABORATIONS AND THE DIVERSITY OF YOUR PORTFOLIO

Ideally, collaborations allow you to increase your research productivity and conduct experiments that would otherwise be impossible to do on your own. However, in the pre-tenure period at a major research institution, collaborations can be dangerous. The danger comes from two facts. The first is that your tenure committee and letter writers have a tougher time figuring out who was the brain behind the operation. The second is that the success of collaborations is not entirely under your control, so you may spend a lot of time on a collaborative project without getting anything published from that work.

When you start your job, you will have some inherent collaborations with your mentors from graduate school and your postdoc that carry over. Some of the papers have yet to be written or are in progress when you start your new faculty position. These are essential projects to keep going. These papers will sustain you and demonstrate productivity for grant reviewers, tenure letter writers, award committees, and your tenure committee. These papers will be submitted and forthcoming even as you are setting up your equipment and before your lab is really productive. Do not walk away from these projects. For many people going up for tenure, these papers with previous mentors account for one third to one half of the papers on their CV in the tenure packet. This includes people who win early career awards. You can look this up. I did recently and was struck by the just how impactful the publications from people's graduate and postdoctoral work were in fleshing out the CVs of Troland Award winners (the Troland Award is an early career award from the National Academy of Science for experimental psychologists and neuroscientists).

New collaborations can be a bit tricky. In general I am warning you to put these off until your own independent line of research is well underway. The economics concept of opportunity costs is applicable to this situation: if a company is engaged in producing some product, that means that it does not have the capacity to produce some new product if that opportunity should present itself. Collaborations can have these same costs. If one of your two graduate students is working on a project with someone in your department, or another department, or from another institution, then half of your personnel is tied up in the collaboration. This is a good if the collaboration pays off with a major research grant and high-impact papers. However, it will seem like squandered resources if the collaboration does not yield interpretable results or never pays off in terms of publications and funding. So how do you know if you should invest resources in a collaborative project?

A good rule of thumb is to ask yourself if you have three or five projects going on in your laboratory without the potential collaboration. That is, are each of your two graduate students running experiments for two or three lines of work that will be written up with you as the senior author? If so, then you can be confident that you will have enough independent work so that your tenure packet will be strong even without this new collaboration. You may recall that approximately 3 papers per year during your pre-tenure period is sufficient for tenure at most major research institutions. In an animal lab, it might be two lines of work recording from animals during a type of task. Why do you need more than one line of work? Some will not produce publishable results even after initially encouraging results. This is why I recommend having more than one project in the works before you take on a new collaboration. In summary, if you have enough work coming out from your own laboratory to demonstrate that you have a thriving independent research program, then go for a potentially impactful collaboration with another lab. If not, then politely tell the potential collaborator that you are really excited about the idea, but that you need to get a couple more projects going in your lab before you will be free to collaborate.

A collaboration likely means having both you and one of your trainees spend time on the collaboration. If you assign a graduate student work on the collaborative project, also have them work on a project that is just with you. That way, if things beyond your control derail the collaborative project, the student isn't in a position of having nothing to show for their months of effort. Remember that your students need several papers per year by the time they are finishing graduate school to secure a good postdoc or faculty position. So having them do something safe under your guidance is a good idea in case your collaboration falls apart.

This is a good time to talk about taking chances with projects in general. Collaborations can be risky. However, many of us also have a variety of projects in our own labs that span the scale from low risk (almost certain to work) all the way up to high risk (unlikely to work, but it would be huge news if it did). Most people like to have a diverse portfolio. If you have all low-risk, low-reward projects in the pipeline, then you are unlikely to have a large impact in your field. This can make it more difficult to get funding, high-impact publications, and strong letters from peers when it is time to put together your tenure packet. Also, this can lead you to become bored with your own research. Being bored by your own science is the kiss of death for the drive you will need to keep things going in your laboratory. In contrast, if you have all your personnel working on the next cold fusion project (you should look this up online if you are not familiar), many of the projects are likely to fail and you are unlikely to produce enough to support successful grant proposals and get tenure. Because of this, I recommend maintaining a diverse portfolio of projects. Collaborations, while sometimes risky, can be good so that you can do novel, high-impact research. However, you will also need some follow-ups on previous work in which you sort out the details of your previous publications or examine phenomena initially reported by others. By keeping a broad portfolio for yourself and for your individual trainees, you can keep your path toward tenure on track even when your high-risk science doesn't pay off.

WHY YOU SHOULD AVOID SPENDING ALL YOUR TIME ON YOUR CLASSES

The heading above is a test. If you have been reading each chapter of this book in a linear fashion, you know that we gave the same advice in Chapter 8 on getting tenure at a liberal arts school. You can imagine that if you are in a position in which at least 50% of your time is supposed to be dedicated to research, then you really don't want teaching to suck up all your time. It is easy to have teaching become all consuming. You can always prepare more for your lectures, and students would almost always like more access and attention.

One concrete tip that I got from a senior colleague was to schedule classes at the beginning of your day. For most human beings this is the morning. The reason is that it is difficult to avoid working on your lecture when you know it is coming up. If your class is scheduled for 2 p.m., it is difficult to resist opening your lecture slides at 8 a.m. and fine-tune them for the next four hours. No one wants to give a boring or poorly prepared lecture in front of 200 eager young people. However, if you prepare your lecture slides for 30–60 minutes the day or night before a lecture and give that lecture first thing in the morning, you will do a good job and be engaging. You might be bright enough to win a teaching award with this level of preparation. If doing your absolute best on a class that you've taught before requires several hours each day, then even winning a teaching award is not going to offset the negative impact it will have on your research output. By teaching at the beginning of your day, you can turn to writing and research by midmorning and work on building a CV that will get you tenure.

Please note that if you are teaching a class for the first time, known as a *new prep*, it will probably require many hours. Typically, when people take on a new prep it does take several hours to prepare for each lecture. For this reason, you want to prep a new class during one of your first semesters and then teach that class over and over again. By the second or third time through the class, you will have to invest much less time in getting ready each day. By then you will also be ramping up the writing of papers coming out of your lab.

Teaching undergraduate classes can be very difficult at times. I find that my undergraduates ask for things that they did not earn. I cannot count how many meetings I have had with students telling me how hard they worked to learn the material for the test that they did poorly on. They frequently ask if they can get some more points. You will have to get comfortable saying no. I always appeal to equity. I tell them that I would have to give extra points to everyone who worked hard, but performed poorly, if I give it to them. Sometimes I appeal to the value of their degree, saying that they don't want a degree from a university that just hands out A's. They would not like this any more than buying a car from a company that produces crappy cars. However, some students have learned that they can just constantly annoy people and wear them down. I once had a student complain to me, my chair, and my dean about a grade from the spring semester all the way through the summer until right before the fall semester started. Usually such stories are about how they dreamed about being a doctor and that I destroyed their dream. Don't fall for these guilt trips. Keep in mind that you are helping them learn about how the world works. When your students leave school, they will need to get a job or go into an advanced degree program. Neither path tolerates incompetence. If students can learn that they need to show up and do their best every day in college, then their degree will have been worth it, regardless of their GPA.

Some of my most energizing experiences came from teaching students who were keen and thinking through the material. These kids helped me realize the logic inherent in the science. Your students will sometimes ask questions that can only be answered by experiments that have yet to be done and are a good idea to run. Although the caution here is not to spend all your time improving your classes, do keep in mind that you are an educator. Quality teaching typically helps you think more clearly about your own research. In addition, teaching can communicate your passion for your science and help attract the next generation to your field of study.

STAY THE COURSE

We have discussed some of the new obstacles that you will tackle and many of the ways to keep fine-tuning the skills that got you a job. I'll reiterate that most of what I needed to do to get tenure was to just keep doing what I had been doing. The hard part was continuing to suppress those thoughts that my work might not be enough, regardless of what I did, and not imagine catastrophes. But the best advice I got as an assistant professor was to work on the experiment or document right in front of me. By taking the next, simple, concrete step, I could confidently walk into the next phase of my career.

Ashleigh's Interview with Timothy P. McNamara, PhD

vanderbilt.edu/psychological_sciences/bio/timothy-mcnamara

Timothy McNamara, PhD is a Professor of Psychology at Vanderbilt University. His research focuses on understanding how human memory for spatial relations guides navigation and behavior. Tim also study mechanisms of memory retrieval using priming paradigms. Tim has been a model of service at Vanderbilt University, serving as Department Chair for 8 years before joining the upper-level administration for 12 years as Associate Provost and then Vice Provost for Faculty (his portfolio also included internal affairs for most of that time). Tim's work at the institution level, overseeing tenure and promotion cases, provides him with a unique and valuable perspective on the tenure process.

DO THE HOMEWORK

Tim encouraged faculty to do their homework to understand the standards and expectations for tenure and promotions at their institution in terms of research, teaching, and service. Tim pointed out that these vary in their importance depending on the individual university. He also recommended looking up the statements on tenure policy and promotion guidelines at your institution. If you feel that the policy has vague points, you can talk to other faculty in the department, the chair, or the dean about the policy. After he said it, this seemed like an obvious answer to the question posed in the title of this chapter.

Tim then discussed how the system works at Vanderbilt to provide a concrete example of the tenure process from an administrative perspective. At Vanderbilt, successful candidates for tenure and promotion are expected to demonstrate research excellence, highly effective teaching, and satisfactory service. The strength of the adjectives (excellence, highly effective, and satisfactory) provides a ranking that helps applicants know how to manage their time across research, teaching, and service. For example, given that the expectations for service are not as high as for research and teaching, faculty members should spend less time on service and more time on research and teaching. However, unlike at some R1s, bad teachers will not get tenure at Vanderbilt no matter how impressive their research is. In addition, problematic teaching can cause faculty to not get contract extensions at pre-tenure reviews, preventing them from even getting to the tenure vote stage.

At Vanderbilt, once the department makes a decision on a tenure application, the candidate does not hear anything until the final decision is reached by the administration months after the departmental vote. Tim feels that this can make junior faculty feel as though the process beyond the department level is secretive. However, when he was in the provost's office, Tim gave a promotion and tenure orientation for all newly hired tenure-track faculty. In his presentation, he went over the tenure and promotion process in great detail, including the expectations, content of the tenure dossier, various aspects of evaluation, stages of the decision, who makes the decision at each stage, what the options are if the evaluation is negative, and so on. Thus, Tim argues that the process is laid out in detail in the faculty manual and elaborated in PowerPoint slides posted online and the applicant should not be ill informed. Although knowing about the tenure process is not going to help an applicant get tenure, it should reduce anxiety, especially knowing that the process is not random and capricious but that real people are reading the final materials and understand the decision they are making is very important.

FIRST PRE-TENURE REVIEW

Tim suggested that applicants use any pre-tenure reviews to help them gauge their progress toward tenure. Vanderbilt has two pre-tenure reviews, after which faculty members may or may not receive an interim appointment. The first pre-tenure review is less informed by data because faculty members may not have had the chance to be particularly productive yet; perhaps they have taught only one or two courses and are still setting up their lab. It is possible to not be reappointed after this review in extreme cases (e.g., really bad teaching), although it is rare. Tim noted that any faculty members who were looking over their teaching evaluations or talking to their chair at least once per year would be aware of the kind of problems that would trip someone up at this early review. In most cases, faculty members receive another appointment that moves them closer to the tenure vote (at Vanderbilt this is a two-year contract).

SECOND PRE-TENURE REVIEW

The second pre-tenure review is much more substantive and serious. This is the review during which a significant number of faculty may not be reappointed. The reason people do not receive another interim tenure-track appointment is usually that they are not making sufficient progress in their research (e.g., grants, publications) or for really poor teaching. Tim said you don't have to have a grant at this point, but if your research depends on grant funding, you must have submitted a grant and shown evidence that you received promising reviews, indicating that the research is potentially fundable.

The pre-tenure focus on grants is really to make sure the faculty member is getting external support for their research, instead of it benefiting the institution. The university does not actually care about grants per se because they actually tend lose money on grants;[1] even government-funded research has to be supplemented with institutional funds (e.g., tuition dollars, endowment returns) because indirect costs paid by the government do not completely cover the true costs incurred by the university. Since the university does not profit from government-funded research, the sole reason faculty need to get grants is to support their research. In fields like philosophy where you can be productive and impactful without a grant, grant funding is not required for tenure and promotion. The real reason the university expects you to get external funding is because they want you to be productive and impactful in your field. In many areas of psychology research, having a large impact requires money to fund essential things like staffing the lab, paying subjects, buying equipment, and so on.

---

1 https://www.nature.com/news/indirect-costs-keeping-the-lights-on-1.16376

In terms of teaching, the advice Tim gives tenure-track faculty is to be in the *good* range on student evaluations, assuming that this information is collected at the candidate's institution. At Vanderbilt, the evaluations are on a scale from 1–5 and the good range, which is a level of performance that won't raise questions, is about a four. If you're below a four on average you should find out why. For certain classes, the chair can provide useful context. For example, in statistics classes the historical mean rating for the class is likely low, and your rating of three might be fine in such a context. You should read and address any comments the students bring up in their evaluations. The comments can be big red flags for the people who read your file on tenure review committees, but they are often easy to resolve. For example, problematic complaints that Tim has seen include faculty members being rude, selectively talking to either only the men or only the women in class, being unorganized, showing favoritism, not getting graded work back in a reasonable amount of time, being unresponsive, or not showing up for appointments. These are easy to problems to solve; you do not need to be a talented lecturer to be polite, fair, answer questions, and so on.

With respect to service, Tim said that there have been extreme circumstances where a faculty member didn't get tenure based on their service. This would occur if the candidate was extremely difficult (e.g., being uncooperative, uncivil, not responding to a chair's request to engage in relatively minimal service activities). Tim advised that with respect to service, just do not be a jerk. If the department chair asks you to do something, just do it if you can. It takes a lot of hands to run a department and it's good to pitch in. However, it's very easy for junior faculty to get sucked up into service because it's easier than writing or coming up with creative ideas, so Tim cautioned you to be mindful of how much time you put into service.

Tim said that when he was in the provost's office, if a faculty member passed the fourth-year review, there was about an 80% chance of being tenured. Although there was some chance of being turned down, it was low. He said that this is partly because a positive vote at the fourth-year review conveys that if faculty keep doing what they are doing, they will be well positioned for a positive tenure decision.

TEACHING EXPECTATIONS AT AN R1

One of the benefits of inquiring about teaching expectations is that you will find out about unusual expectations among a given university's peers and aspirational peers. For example, it is imperative that junior faculty learn how important teaching is for tenure at Vanderbilt. This might be unusual among R1s where teaching may not be as heavily weighted.

Tim acknowledged that because getting a job at Vanderbilt requires evidence of extensive research experience, new faculty may have little to no recent teaching experience from their postdoc. To develop into the high-quality teacher that Vanderbilt expects, Tim recommended that candidates consult the Vanderbilt Center for Teaching or the comparable version on their own campus. Tim has found that the Vanderbilt Center for Teaching uses very effective methods to teach pedagogy and mentoring skills, such as video recording lectures, conducting confidential midterm evaluations, and hosting workshops. Although it is common that new faculty might struggle in the classroom because evidence of teaching effectiveness is not required to get hired, eventual improvement in teaching is essential. Tim often refers junior faculty who are struggling or who just wish to do better in the classroom to the Center for Teaching. During the tenure process, just making the effort to improve by seeking out these resources is a positive mark. The expectation is not that every faculty member turn themselves into a master teacher. However, Vanderbilt invests a considerable amount of money in the Center for Teaching with the expectation that junior, and even senior faculty, will use its resources to navigate the demands of teaching and become ever better educators. Therefore, the worst thing that can happen with respect to teaching skills is when a junior faculty member who gets poor teaching evaluations clearly hasn't sought out any resources.

WHAT ELSE JUNIOR FACULTY CAN DO BESIDES PUBLISH

Tim listed a number of things junior faculty should do on the road to tenure. First, they should meet regularly with their department chair or dean to get a sense of how they are progressing. These meetings should happen at least once a year. Many departments institutionalize this by having an annual review or a faculty mentor, but plenty do not, so faculty members should establish their own regular meetings if necessary. The purpose of going to the dean's office is obtaining a college-wide, institutional perspective on their progress. The department chair will provide information about the expectations of department colleagues, but after the department, a dean is also going to make a decision on tenure and promotion. So meeting with someone in the dean's office to understand their perspective is useful and might only require one meeting during the entire pre-tenure period. It probably will not be the dean of your college that meets with you, but rather an associate or assistant dean who is responsible for meeting with and talking to junior faculty about the institution, what the expectations are, and so on.

Second, it is important to find mentors in senior colleagues in the department to learn about the institutional culture and the department's expectations. Knowing these can help you navigate the promotion and tenure process. Beyond one's local environment, Tim pointed out that candidates who are hired at R1s have inevitably been surrounded by very successful people in the field (e.g., graduate and postdoc advisors). He suggested looking at their careers and what they did and to the extent that it fits, trying to model your own career after theirs to learn from what they've been exposed to as they built their own careers.

Third, faculty can really get caught up in service. This can be dangerous in that it takes up your time and prevents you from focusing on the more important tasks of research and teaching. This is particularly problematic for junior faculty in underrepresented groups, such as women and minorities. Why? Because when underrepresented faculty are hired, they can be bombarded with requests to serve on committees so the committee can be more diverse.

Tim warned that women and faculty from minority groups could spend all their time serving on departmental, college, and university committees. It is imperative to resist this temptation. It can feel like people are really pleased with your work and you are making a great impact on the institution. However, when it comes time for tenure and promotion, little of that committee work is really going to matter much. Rather, what people are mostly going to be looking at is scholarship and teaching.

Ultimately, it's really important to serve wisely. In particular, Tim advises candidates that if they feel like they are being buried or torn apart by service demands, they should talk to their department chair or head and ask them to intervene on their behalf. For example, ask the chair to tell someone making a service request that the faculty member is new, and the department doesn't want her or him to spend too much time on service, so to please find somebody else. Although there is no rule of thumb for how much service is too much, Tim advised faculty to find out if they are doing too much by talking to the chair or senior colleagues. This is an example of when having that senior mentor in the department would really help because they can assess your situation without a vested interest in you doing more service.

Fourth, use all the resources that are available. For example, all R1s have some kind of center for teaching and learning. Tim advised taking advantage of programs that help junior faculty. In addition, he suggested taking advantage of opportunities to self-promote, such as networking with visiting scholars. When Tim was pre-tenure, every time he finished a paper and submitted it, he would also mail a copy to every senior person in his research area. His reasoning was that they probably wouldn't read it, but if they didn't have it, they definitely wouldn't read it. In specific instances, this helped Tim increase his reputation in the field. These days people often email a link or a pdf of the paper.

Fifth, spend your startup funds. The purpose of startup funds is to help you get going and get tenure. So spend them! Tim pointed out that spending startup funds is important in leveraging the best shot of getting tenure. Obviously, the funds won't do you any good later if you don't get tenure. Ideally you should budget to run out of money the day you get tenure. This advice is particularly targeted toward female candidates. In Tim's experience, female candidates can be especially conservative with their startup funds, and his observation is consistent with data on gender differences in risk tolerance for financial investments). Tim can understand why candidates might decide to be financially conservative. Perhaps they don't want to run out of money. Or they might want to protect themselves against not getting external funding. But if getting a grant is important to getting tenure in your field, then having the bulk of your startup funds left over after tenure won't be very useful. If you successfully secure external funding, then maybe you'll have a ton left over. But if you don't get a grant, you probably won't get tenure, and you can't take your startup funds with you. Again, Tim's bottom line recommendation is that people budget their startup funds so they use the money to advance their research and are not sitting on a pile of money at the time of their tenure decision.

WHY WOULD THE DEPARTMENT VOTE YES BUT THE DEAN VOTE NO?

Tim said that one main reason the department may vote in favor of tenure but it doesn't get approved at a higher administrative level is that the department is not looking at the tenure packet objectively. By the time someone comes up for tenure they are usually a beloved member of the department and the department is a human institution. Department colleagues know candidates personally. They have spent years having meals with them, probably met their family, and so on. This means that colleagues in the department are naturally emotional about voting in tenure situations. Tim believes that many departments tend to give the tenure candidate the benefit of the doubt and vote in favor of a case. Tim said that it is somewhat common following a split department vote for tenure (i.e., some people will vote yes and some vote no) for the tenure case to get turned down at the dean's office.

DISILLUSIONMENT

Occasionally, junior faculty become disillusioned with how they spend their time. Tim thinks this frustration may come from not clearly understanding the department's expectations. He suggested meeting with the department chair to find out what the priorities are for tenure and using that information to help allocate time.

If the department chair is not helpful, it is a good time to consult those senior mentors in the department that Tim previously advised leaning on when needed.

Sometimes new faculty feel like they are nothing more than grant writing machines, especially in medical schools, and like they cannot spend as much time in the lab or thinking creatively as they would like. Tim's advice to faculty who feel this way was to think about what really makes them feel passionately about their science and try to spend more time doing it. He suggested carving out a little more time to spend on their passions—just making it happen at the expense of other things that they don't enjoy doing. Prioritizing interests will further your career because getting absorbed into administrative work, such as serving on committees, might hurt overall productivity and negatively impact tenure.

PROMOTION OF WOMEN AND OTHER MINORITIES

Although Tim has never experienced gender disparity in salary and startup package negotiations, he realizes that such differences have been documented. On one hand, Tim said, it is up to junior faculty to solve that problem by making sure that they are asking for what they need and without being embarrassed about it. It would be wise for junior people, particularly women and minorities, to get advice about whether what they are asking for is adequate from a trusted mentor or advisor.

The imbalance of gender and minorities at higher ranks (i.e., associate and **full professor**) is apparent at every university and is confounded by generation. Tim pointed out that in many fields there were fewer women and minorities applying for these positions 20–30 years ago. Given that the junior ranks (i.e., assistant professor) are much more balanced now, the expectation is that over time higher ranks will be balanced as well. However, Tim acknowledged that it does take both a top-down and a bottom-up effort to change the disparities. In terms of promoting gender and minority equality from the top, there has to be an institutional commitment to sending a consistent message across levels of administration. Department chairs, deans, and provosts all need to be committed to diversity and gender equity.

At Vanderbilt, this has sometimes required the provost to lean directly on department chairs during the hiring process and basically say, "I'm not satisfied with you bringing to us yet another white male candidate. I want to see who the other candidates were," or "I want you to bring us a more diverse slate of possible people to fill this position." In Tim's experience, the provost's office has been very involved (with good intention and execution) in getting schools and departments to diversify their hiring.

Tim said that every university claims that they are taking steps to increase diversity, but not all really are. The best way to really know if they are being proactive is to look at who gets hired at an institution. Tim explained that Vanderbilt endeavors to hire people who have the skills and abilities to succeed, so increasing diversity is not a matter of changing requirements or expectations.

One method of increasing diversity is broadening the slate of what the department is interested in hiring for. Sometimes this means that the search committee needs to broaden their criteria by considering a slightly different research portfolio. In one example, Tim said that some departments claimed for years that they couldn't find any African American faculty members who fit their job descriptions. When Vanderbilt hired a new energetic head of African American and Diaspora studies, this person went out and found really interesting, talented scholars who met all the quality requirements of any leading department, but their research areas were a bit outside the narrow boundaries established by the departments. It took someone outside the department who was able to think creatively and imagine a scholar who wasn't narrowly bound to a traditional research area, but they were able to find spectacular African American scholars.

THE GOAL ISN'T ACTUALLY TENURE

Although this chapter is about promotion and tenure, Tim advised junior faculty that what they really need to focus on is career development, rather than tenure. Vanderbilt does not hire people because they think they can get tenure. Rather they hire people because they think they will be outstanding in their field. Tim echoed Geoff's implicit sentiment in the chapter that staying focused on being successful in the field will translate into tenure and promotion. Getting hung up on promotion and tenure is a mistake. What faculty really need to focus on is their long-term career development and becoming a star in their field. If they do that, they will get tenure at Vanderbilt or any R1 institution.

CHAPTER 10

How to Get a Grant
by Geoff & Ashleigh

Any book about navigating a research career must talk about obtaining funding for your research. Getting grants has become an increasingly difficult part of a scientist's job. Several decades ago, the top 30–40% of grants were funded by the NIH and the National Science Foundation (NSF) of the United States.[1] These are the two biggest, and most important funders of research in psychology and neuroscience in the United States. European countries and the European Union (EU) have similar agencies that fund research. These days, the funding percentages at NIH and NSF are 5–20%, depending on the specific institute or division that your application is assigned to. Researchers in the United States are not alone in chasing ever more elusive grant funding. Restructuring of the EU, financial recessions, and political instability across the globe have made it more difficult for us scientists to secure funding for scientific research.

The good news is that there are some simple things that you can do to improve your odds of writing a grant that will fund your work. We are by no means masters of the universe when it comes to getting grant funding. However, we can share what we have learned. As described below, much of what we have learned is how to align the grants we write with our research experience and expertise. However, we have also learned through experience that we all lose sometimes. Tolerating and accepting that a grant you are very excited about might not get funded is part of the game. But we can give you tips for how to increase your odds of winning.

We begin by discussing general principles aimed at getting a major research grant from a governmental funding agency. We then turn to how to get smaller grants from professional societies.[2] In the sections that follow, we describe principles that we believe are generally useful to help you get a grant.

---

1   https://nexus.od.nih.gov/all/2015/06/29/what-are-the-chances-of-getting-funded/, https://www.nsf.gov/attachments/134636/public/proposal_success_rates_aaac_final.pdf

2   Not all large grants are from governmental agencies. For example, the Howard Hughes Medical Institute makes large awards to fund basic and applied research for a few investigators; https://www.hhmi.org/about/policies/funding-policies

TO WHOM WOULD YOU WANT TO GIVE A GRANT?

One of the most useful things to help you write good grants is knowledge about how grants are reviewed. The simple answer is that grants are reviewed by us, the scientists. It is very much like the process of peer-reviewing papers, except that after writing their reviews, the reviewers, known as *study section* members, meet to make a recommendation for funding to the institute. Another exception is that members of study sections are even more risk-averse than paper reviewers. So to put together the best possible application, you need to think about what you would need to be convinced of by the grant.

As we discuss at length in this chapter, the reviewers decide whether your project is worth spending about $1,000,000 USD of taxpayer money, assuming it is a typical grant at an agency like the NIH; the typical grant at NSF comes in a bit below that. As the writer of your grant, you are called the **Principle Investigator (PI)**, and your goal is to convince three to five experts in your field who are on the study section reviewing your grant, that your project is a safe investment for taxpayers and will move the field forward.

First of all, the reviewers of your grant are going to want to know that you are an expert in what you plan to study. They also want to know that you know how to use the tools and methods to study what you plan to study. This is the first thing we mention because it is something that you need to think about long before the grant deadline approaches. You typically want to position yourself for your grants years in advance. It can take years because you want a publication track record that can convince reviewers that the grant money is a good investment. Would you want a heart surgeon to do your surgery if she had been doing foot surgery, but recently realized that it would be easier to get a big grant if she were a heart surgeon? Of course not. Similarly, you should not submit a grant without demonstrating that you already know what you are doing when it comes to the topic of study and the tools you propose to use to understand that topic.

If you have been using a method for several years and have a good publication record proving it, then it can be possible to pivot to a new question or topic of study pretty quickly. It might only take one or two publications on that topic or a collaborator with the right expertise. Conversely, if you have already established your expertise in the study of a topic, and then want to add a new method to that study, this might only take a paper or two, or some very convincing preliminary data, to sufficiently convince the study section of your ability to pull off the research. The mistake is to reinvent yourself in the 6–12 pages of your grant proposal.

Why do we bring up the idea that someone might feel compelled to reinvent themselves in their grant? What happens to many researchers is that they don't get the first grant that they submit. Many don't get the first 5–10 grants they submit. At this point, they panic and think that they need to reinvent themselves and submit grants on different topics or using different methods. This is the wrong approach. Grants are no place to reinvent yourself. But you may think that you have to because you need to submit grants that are completely different from the unsuccessful grants that you've been submitting.

But those completely different grants are usually not backed up by publications showing that you are one of the world's experts on that topic. What we have found through our own experience, and that of many colleagues, collaborators, and advisees, is that the successful strategy is to stay the course. Keep publishing on the main line of study in which you have already established excellence.

If you keep building a coherent line of research, ultimately your CV (or biosketch, as it is known in funding agency jargon) becomes undeniable. If you want to add an exciting new wrinkle, perhaps you add a new method that you have published a recent paper on, or you could add a collaborator (known as a co-Principle Investigator, or co-PI) to the grant to establish that you have someone with the credentials to pull off the proposed research that goes beyond your expertise. The buzzword in this regard is *feasibility*. The reviewers need to know that it is feasible for you to conduct the proposed research.

Typically, you demonstrate feasibility not only with your publication record, but also by providing *preliminary data* in the grant. Preliminary data is the grant term for pilot data. These days this has become so essential that people are expected to provide pilot data for each major theme in the grant. If you propose to do something new and exciting, you need to show that you can. This is usually done by showing a figure of the results obtained from pilot experiments. You can then propose additional experiments that replicate, extend, or add novel twists on this general theme.

This emphasizes the importance of being able to run your laboratory for a couple of years with the startup money that your institution provided as a result of your startup negotiation. You can now see how important it is to have operating costs included in your startup package; you will need to conduct your pilot studies before having the benefit of your first major research grant. You may also remember from Chapters 4 and 5 that when we discussed graduate school and postdoctoral training, we recommended that you think about what your first R01 grant would be, and start collecting preliminary data even before you have your faculty position. You can use preliminary data collected years before your grant submission. However, study sections need to see in your grant submission that data can be collected in your new laboratory. Typically an efficient way of showing this is by explicitly stating in the figure captions that the preliminary data shown were collected at your current institution.

So how much preliminary data do you need, and is it possible to have too much? The short answer to the second part of that question is that it's probably not possible to have too much pilot data. For example, in the last ten years we have had the experience of submitting a major grant that was criticized for not having pilot data for the future directions proposed at the end of the grant! Clearly this is an outlier. Our rule of thumb is to provide preliminary data for each major junction of the grant. If we use one task in Aim 1 (Aims are the major goals of a grant), and different tasks in Aim 2 and 3, then we need to show that we can analyze and interpret the data from all of those types of tasks. You need to show that you can use each piece of equipment or use each method that you propose in the grant. Sometimes prior publications can serve this purpose. Obviously, you can't have published the research that the grant proposes to conduct, but it is not clear that you can overdo the presentation of data that shows readers that you can do what you propose.

Now, what is the minimum amount of preliminary data? This varies based on the group of reviewers you get, as standards vary across different scientists. It is safe to assume that each aim of your grant will need some preliminary data to demonstrate feasibility. Typically, NSF and NIH grants have approximately three aims. Sometimes, two big aims work. Sometimes, four highly related aims will work. This means that you might need to get three or four experiments running and collect a significant amount of data prior to submission of the grant. We have found that it typically takes months if you are starting from zero.

In sum, reviewers want to fund research that they know will work. This is where preliminary data comes in. The other strategy you can adopt is to propose experiments that will be useful regardless of the results and describe how the different possible outcomes will distinguish between competing models or theoretical perspectives. It is not wise to submit a grant in which the experiments will provide confirmatory evidence for a hypothesis only if they turn out one particular way. Instead, you need to describe how the different patterns of results will be useful.

HOW TO WRITE THE PROPOSAL

This section is specifically about writing grants, but also generally about scientific writing. A grant proposal is a 12- to 15-page document (sometimes as little as 3 pages, such as for certain private foundation grants) in which you try to convince the reader that you need 3–5 years of funding. It is like writing a couple of *Science* or *Nature* papers that will determine your future. No pressure. The style is pretty similar to that analogy. We mean that you really want the figures and the text of your grant to be crisp, clear, and aesthetically pleasing.

One of our former mentors described getting a grant as winning a beauty contest (one in which individuals of all sexual orientations and genders are treated like pieces of meat). You should keep this in mind when writing your proposals and working on the figures. You want the parts of the grant to be pleasing to the eye and fit together in a way that is slick, clean, and convincing. When working on the figures that you insert into the grant, you should use *Nature* or *Science* style so that the reviewers can be reminded that you are aiming for your research to end up in the highest impact journals possible. By having your figures look like those in the glossy magazines, you show your reviewers that you know how to play the game at the highest level. Remember to minimize white space and have the figures work to communicate ideas or show findings that would take too many words. On the following page is an example of what we are talking about from one of our grants, in which we used a provocative image in the specific aims to catch the reviewers' eye and make the image memorable. Opinions vary about including such an image in your specific aims, but we hold a high opinion of this trick.

Aesthetics are important, but clarity is king! Propose a hypothesis that can be simply stated in one sentence. Then propose a competing hypothesis that makes different predictions for your experimental design. In the proposal, state that hypothesis X predicts A and hypothesis Y predicts B. Then, with several experiments, distinguish between these competing hypotheses. The three aims of your grant might use three different types of evidence to test these two hypotheses with different methods to provide converging evidence for the conclusions. Or you might have hypotheses that layer on top of each other such that each aim tests more refined versions of the hypotheses. Ruling out a hypothesis definitively would have significant theoretical impact that would move the field forward, assuming that the hypothesis follows from existing theories in the field. This may sound elementary, but you would be amazed at how many grants get shot down because the investigators fail to discuss alternative outcomes of the experiments proposed or the theoretical significance of the proposed work.

The best way to write clearly is to avoid jargon and explain things simply. You might believe that science is complicated and that you will sound like a brilliant scientist in your proposal if you write dense, complicated sentences and long paragraphs packed with ideas. You might also think that people will think you are smart if your writing is more complex. You would be wrong.

One of our favorite papers about writing showed that when researchers gave readers two passages to read. One version had simple words and the other version had complicated words. People judged authors who used simple words as smarter than authors who wrote with long, fancy words.[3] This is the hidden payoff of writing simply. Not only will the reviewers understand the experiments you are proposing, but they will think that you are smarter than if you wrote the same thing in a complicated, jargon-filled way.

Another mistake that writers all make is to overestimate their audience. People look at study section rosters filled with top researchers in psychology and neuroscience and think that they need to impress them with a dense proposal filled with intricate, complicated methods to show that they are worth the years of funding. Researchers on study sections are actually overworked academics who review papers for free and are trying to keep their own publication rates high. They have agreed to review 7–12 grants for the study section on top of all their other responsibilities. Imagine adding three additional reams of paper to your annual reading load. For this reason, you need to write proposals

---

3   Oppenheimer, D. M. (2006). Consequences of erudite vernacular utilized irrespective of necessity: Problems with using long words needlessly. *Applied Cognitive Psychology, 20,* 139–156. doi.org/10.1002/acp.1178

SPECIFIC AIMS

Schizophrenia is a debilitating disorder that affects more than 51 million people worldwide. Although the familiar psychotic symptoms of schizophrenia are commonly known, it is less well known that the cognitive deficits drive much of the disability, placing sharp limits on social and occupational success for patients (Green, 1996). Research suggests that an inability to regulate behavior by adaptive control may be a core feature of schizophrenia (Malenka et al., 1982; Lesh et al., 2011). Adaptive control refers to the ability to override routines and habits, and produce contextually appropriate behavior. Adaptive control is hypothesized to be implemented by the low frequency oscillations emanating from the medial-frontal cortex to coordinate activity across the diverse set of brain areas required to perform any given task (Luu et al., 2003; Wang et al., 2005; Cavanagh et al., 2009; van Driel et al., 2012; Anguera et al., 2013; Narayanan et al., 2013; Cavanagh and Frank, 2014; Cohen, 2014). Medial-frontal theta (4 – 8 Hz) oscillations measured with the electroencephalogram (EEG) appear to be elicited by situations calling for increased control of cognitive processing across a variety of situations, including stimulus novelty, response conflict, negative feedback, and behavioral errors (Cavanagh and Frank, 2014). Moreover, these theta oscillations have been hypothesized to serve as the carrier signal that coordinates neuronal populations involved in implementing control (Cavanagh et al., 2009; van Driel et al., 2012; Narayanan et al., 2013; Cavanagh and Frank, 2014), with medial-frontal cortex working in concert with prefrontal and sensory areas to support flexible, adaptive behavior (Ridderinkhof et al., 2004; Narayanan and Laubach, 2006; Sheth et al., 2012; Narayanan et al., 2013; Bonini et al., 2014). For example, when an error occurs, network-level oscillations allow executive mechanisms to adjust subordinate cognitive processes (e.g., perceptual attention or response-selection thresholds).

In this project we will record EEG oscillations from patients with schizophrenia and healthy controls (as previewed in **Fig. 1**). Our preliminary work indicates that abnormal medial-frontal theta oscillations may underlie impaired adaptive control in schizophrenia. Next, we will determine whether noninvasive electrical stimulation that normalizes theta-band dynamics can restore adaptive control in schizophrenia, as the theta-band hypothesis of adaptive control predicts. Our preliminary data suggest that our stimulation protocol can successfully restore inter-regional, frequency coupled network activity in the patients (Uhlhaas and Singer, 2010) by phase aligning low frequency oscillations (Ford and Mathalon, 2008; Lesh et al., 2011). Thus, this project will establish a causal link between synchronized low-frequency oscillations and adaptive control dysfunctions in schizophrenia. Our translational goal is to provide the groundwork necessary to develop a therapy for improving adaptive control and the accompanying cognitive deficits in schizophrenia. The current project will answer three questions in the specific aims.

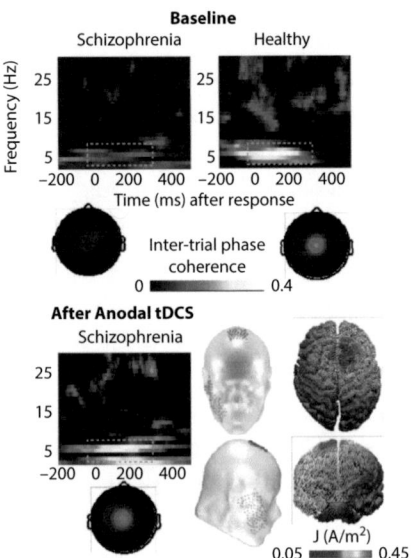

FIGURE 1 Transcranial direct-current stimulation (tDCS) model and spectral signatures of adaptive control (error minus correct trial activity) that are abnormal in patients relative to controls and restored with anodal tDCS of medial-frontal cortex.

Aim 1) Does transcranial direct-current stimulation (tDCS) cause phase realignment of abnormal theta-band oscillations observed in patients with schizophrenia, improving behavioral measures of adaptive control (Exp. 1, preliminary data included)?

Aim 2) Does improved phase alignment of local theta oscillations lead to long-range synchrony between brain regions, allowing for more effective adaptive control of cognitive processing (Exp. 1–2, preliminary data included)?

Aim 3) How general are the cognitive improvements following the restoration of theta band oscillatory dynamics in patients with schizophrenia (Exp. 3–4, preliminary data included)? Figure 1[4]

---

[4] For the published work, see: Reinhart, R. M., Zhu, J., Park, S., & Woodman, G. F. (2015). Synchronizing theta oscillations with direct-current stimulation strengthens adaptive control in the human brain. *Proceedings of the National Academy of Sciences, 122*, 9448-9453.

FIGURE 10.1

that are easy to understand. Try to remember this axiom from communication theory: Less is more. When you try to pack too much in, your reviewers get lost and fail to understand the impact of your work.

Use figures to communicate the experimental design and show the equipment. Have a figure that shows the contrasting predictions of the hypotheses you will test. The first thing reviewers do when they get a grant (or paper) to review is to skim through it and look at the figures. Could a reviewer basically understand your proposal if they just glanced at the figures? If so, that is an asset. It means that in study section, when it has been a couple of weeks since they read your proposal and scored it, they will be able to defend your methods, predictions, and ideas based on the figures that they can glance over during the debate.

One trick that grant writers do is to use a bold font for the important parts of the proposal. This can be the hypotheses and predictions, the novel methods that the proposal employs, and so forth. This is another way to make it easy for the reviewers to quickly apprehend the most critical points of the proposal. Of course, it is possible to take this too far and to end up with a proposal that is all bolded. Be selective, and this can be a very useful rhetorical tool.

Write a proposal that anyone could understand. Could you hand your proposal to your TA or your dentist and have them understand it? You should be able to. The members of a study section have a variety of expertise. For example, the Cognition and Perception study section at NIH may have developmental psychologists, cognitive psychologists, faculty members from medical schools, and researchers who study aging. Use terms that an undergraduate in psychology or neuroscience would understand and then your reviewers will all be happy. This means minimizing jargon. If you use it, define it. Avoid acronyms and abbreviations. No more than three to four such letter strings in an entire proposal. Three to four chunks of information is the average working memory capacity of the average human. Take into account the cognitive limitations of your readers and they will process your proposal efficiently. Better yet, they will think of you as smart.

Do not leave it up to readers to connect the dots and infer the importance of the work you are proposing. You need to be very explicit. Tell them exactly why your work will change how people think about an issue and perhaps how researchers perform experiments. As scientists, we are not very good at self-promotion. If we had been, we would have gotten business degrees and would be selling things or running for the presidency of the United States. Instead, we were trained to be cautious in our conclusions and to sprinkle our declarative sentences with words like *maybe*, *perhaps*, *suggests*, and *indicates*. A grant application is no time for such modesty and self-injurious caution. Be bold, assertive, and lay out a grand vision for your readers. You can revert to being a cautious experimenter in the Potential Problems and Solutions section of the grant.[5]

ENCOURAGEMENT FOR THE JUNIOR FACULTY

Perhaps you are still concerned that when it comes to grant funding, your proposal will be in the same pile as the most famous people in your field. This is true. The same study section will be reviewing grants from the PI that is turning out 50 papers a year and new assistant professors still working their butt off to establish their independence and publish two per year. Even though the same reviewers will see grants from both types of researchers, the funding agencies' program officers look at the scores of Dr. Super-Famous and Dr. Brand-New in different ways. So proceed with confidence if you are junior that you will get the benefit of the doubt if you can get a good score.

RESUBMISSION

So you get the summary statements (this is what the reviews are called) back and your grant does not get a great score or is not scored at all. *Unscored* means that it was not discussed because it was in the bottom half of the grants in the study section. Time to give up, right? No. If you have learned anything about getting papers published, it is that you need to be tenacious. Perhaps the reviewers indicated that you do not have sufficient preliminary data for one of the aims. Collect that data. A resubmission is evaluated on how responsive the PI is to the reviews from the previous round. You also get another page, called the Introduction, to explain the changes you made for the resubmission. By addressing all the concerns of the reviewers, grants can go from unscored to funded.

As with the initial submission, do not rush this back in. No one is measuring latency of response. If you need to miss a cycle (i.e., about 4 months at NIH, 6 months at NSF) to collect additional pilot data or think about control experiments that will address the concerns, that is fine. Make sure you are submitting a grant that you feel is perfect each time. That goes for initial submissions and resubmissions. Proofread until you are sick of it. Then have someone else proofread it a couple of more times.

At NIH, you are allowed to submit a grant twice before it is considered new again. If you do not get a grant funded in the first two submissions, then think about the remaining problems that reviewers identified and how to fix them. One possibility is that they think the approach is simply wrong. Maybe you cannot distinguish between theories using your methods. Or maybe you cannot infer what you proposed to infer from your

---

5   A section starting with this as a header comes at the end of each aim. Tell readers how you would address the problems that might arise with the experiments. This has become a necessary part of the application, in which you show reviewers that you've thought through the pitfalls of what you are proposing. You can often propose control experiments in these sections also.

measures. This kind of fundamental problem is not possible to fix with tenacity. Instead, look at what you have and think about what you can infer. You likely published a paper or two using these methods for this topic, if you followed the advice we provided above. Go back to those papers and move more methodically through the logical space. If you published a paper inferring one implication from a method, then extend that method to the next logical inference. Good science builds on a solid foundation, and you can move back a bit so that you are on solid ground. If you are stuck, ask a colleague in your department or a former mentor to look over the grant and summary statements. They can give you advice about what they would do.

Take advantage of your colleagues. It is normal to give your grant to senior colleagues for comments. Usually new faculty send their grants to their previous mentor, friends in the department, and sometimes even recent members of the study section they think their proposal might be assigned to. You are not burdening people by asking them to look over the grant and give you feedback. Of course, this means finishing your grant up to a month before the due date.

Completing a polished version of your grant early should always be your goal. If you can write your research strategy and then leave it alone for a week or two, it is amazing how fresh your perspective will be when returning to the document. We routinely set a deadline for ourselves that is 1 month before the actual due date for the grant. That way we can pass along the grant to friends for comments, as well as look at the grant with fresh eyes in a week or two.

The other way that you can take advantage of your colleagues is to team up with them. Modern science is team science. If there are questions about your ability to pull off the amazing things you propose to do in your grant, and you have a colleague that has been using these innovative methods already, then bring on a co-PI. The strength of your two CVs together is better than any one person's CV. You cannot just tack anyone on your grant, though. You need your co-PI to have solid credentials for the parts of the proposal in which you are weakest. The fun part is that this can really turn into novel work that each of you might not otherwise do as separate PIs. We are not aware of anyone who would turn down fun new scientific questions along with money, so you should be able to enlist an excellent person to help you. This could also be the beginning of an amazing partnership (think Hubel and Wiesel or Sakmann and Neher).

Reviewers on study sections like it when a grant has a unique aspect to it, and bringing together two people is unique. Occasionally junior faculty members might worry that having a co-PI on a grant is going to hurt them when it is tenure vote time. We have not found this to be the case. Typically, you get your first grant and that gives you the capital you need to get another one without anyone else on it, or with a different co-PI, demonstrating that your success was not simply due to that other person. However, just getting one grant, even with a co-PI, is difficult enough that tenure committees view any grant, on top of your publications each year, as the aspect of your package that seals the deal.

In summary, remember the point we made above about making your CV undeniable. If you continue to publish and do good work, NSF or NIH will eventually want to claim credit for your successes. You may need to get creative and submit a grant under a different 'funding mechanism.' For example, maybe you can submit a grant for an NIH R21 instead of a R01. The R21 has a budget similar to an R01, but it lasts for two years instead of the 5-year maximum of the R01. At NSF it might mean going for a CAREER award instead of the normal NSF basic research grant track (a CAREER award is expressly for early career scientists). This simply involves aiming at a slightly different target. The basic recipe for success is to proceed with confidence and keep building your track record of solid research publications. The publications get you the grant, not the other way around. Later in your career it starts working like a closed loop in which the publications that you get during one grant period set you up for your next grant.

GRANTS FOR PRIMARILY UNDERGRADUATE-SERVING INSTITUTIONS

There are grants for all different career levels and types of institutions. One of the attractions of nontraditional institutions or those that serve a special population might be the select grant funds that are specifically available to faculty at such institutions. For example, NSF has funds that are earmarked for HBCUs. There are also special grants for faculty working at institutions where graduate students are less available. For example, the HBCU-UP at NSF is funding for research taking place at an undergraduate-serving program.

Faculty members conducting research at teaching-focused institutions are eligible for private foundation grants, such as those from Psi Chi, the national honors society in psychology. These grants fund student and faculty research; they fund students to travel to other institutions and conduct research and network; they fund travel grants for conference attendance; and fund putting on your own conference. Another option for faculty at smaller programs would be funding from the APA Summer Undergraduate Research grants.

If you feel out of touch with grant applications because you have spent several years primarily teaching to land a teaching job, we have some suggestions about how to become competitive for these smaller grants. First, you can volunteer to be a grant reviewer via specific organizations' websites, such as Psi Chi. The NSF also invites people to register as Graduate Research Fellowship Program (GRFP) panelists through their website. Serving as a grant reviewer gives you an opportunity to have exposure to different grant applications and firsthand experience of how they are scored and ultimately funded. This is true of manuscripts as well. If you are concerned about getting grant funding because you have not been publishing, signing up as a reviewer at undergraduate-serving journals to get firsthand knowledge of the expectations for submissions may help you improve your publication rate.

Most universities and colleges offer internal grant programs. You should take advantage of these. Sometimes there are small pilot program grants that are often sufficient to get a line of work going and get enough data to populate an external grant proposal. One great way to find out about these is to send the undergraduates working in your lab to the office of research and sponsored programs open house in the beginning of the semester with the specific goal about finding out about potential funding mechanisms on campus. Another avenue to find out about funding on campus is to get the email list for the same office's monthly newsletter.

TALK TO THE PROGRAM OFFICER

Our next piece of advice is applicable for all types of grants (e.g., private foundation, government institutes). The advice is to communicate with the program official. Look up the contact information online and send them an email. See if you can line up a time to chat with them on the phone.

This is good to do before your first submission, particularly if you are unsure whether your research would even be appropriate for funding under their program. For example, it is possible that your research involves recording brain activity to understand basic mechanisms. However, the foundation you want to apply to is focused on funding work to develop computational models of how the brain works. You can tell them about your research and ask if they have advice for how to best pitch this in your application to their organization. Obviously, you should read about the foundation or program ahead of time so that you can try to make this pitch yourself when you call. But the program officer can often tell you about things to avoid saying in your proposal that go against their mandate or how to emphasize aspects of your research that are particularly relevant for their funding priorities. For example, it is possible that they are already funding many grants like the one you were thinking of submitting and that you can spin your work differently to give it an edge when you submit it in the near future. Or the program officer can tell you about essential elements that are included in successful grants for their mechanism. Getting this kind of inside information can make the difference between a successful and unsuccessful grant.

You might feel like you are bugging the program officer and that asking questions will hurt their opinion of you. This is certainly not the case. Our experience is that program officers really want to talk to applicants and feel like they are being misused when not being used.

There is a final type of research funding that we have not discussed in great detail but which is particularly relevant for this section. That category is defense department funding. At the Office of Naval Research, Air Force Office of Scientific Research, and similar Department of Defense research programs, your relationship with the program officer is of the upmost importance. These program officers have great latitude to fund proposals they are interested in and to help you craft successful proposals.

THE TAKE AWAY

We think we all found it intimidating when we began submitting grants. If you have followed the advice in this book, then you submitted grants as an undergraduate, graduate student, and postdoc to fund your own training. These early grants are very useful because their form is very much like that of the major research grants from those same federal agencies. The great news about grant writing is that you will learn and improve your skills each time you do it. Writing grants can improve your ability to communicate the importance of your research and continuing to publish papers will help you write successful grants. Through the process of writing and revising you will acquire skills that will improve your odds of writing a successful grant or writing a high impact paper. The crucible of seeking grant funding often refines your research productivity in unforeseen ways, so we hope you can embrace the process.

Ashleigh's Interview with Mark D'Esposito, MD

despolab.berkeley.edu/despo/

Mark D'Esposito, MD is a Professor of Neuroscience and Psychology at the University of California, Berkeley, where he is also the Director of the Henry H. Wheeler, Jr. Brain Imaging Center. Dr. D'Esposito got his MD from State University of New York (SUNY), Syracuse, and was an Assistant Professor of Neurology at Penn before moving to Berkeley. Mark has trained over 79 graduate students and postdocs. He is the editor of the Journal of Cognitive Neuroscience. Mark's laboratory is currently funded by 4 concurrent R01 grants from the National Institutes of Health and Mark has authored over 325 journal articles.

Mark has impressive expertise in getting NIH grant funding. This is evidenced by his concurrent support from multiple NIH grants. Based on this expertise, Mark focused his thoughts on how to craft grant applications that will be successful at NIH. In addition, we discussed several issues that are particularly relevant for those seeking NIH funding, such as how to portray the clinical relevance of work that is skewed toward basic science. Mark begins by discussing pilot data and then he moves through his thoughts about how to craft specific aspects of your grant proposals.

PILOT DATA

When it comes to including pilot data in your grant, Mark said that there is a fine line between providing too little pilot data and too much pilot data. As a reviewer, Mark has seen grants where the reviewers get a sense that the experiments have already been done, which is clearly problematic. Mark reflected that over his career he has included less and less pilot data in his own grants because the purpose of pilot data is to demonstrate feasibility or that you can get a crucial effect, which often can be done with previous publications. He said that young investigators often forget that previous papers can be used as pilot data, which avoids the problem of seeming as though the experiments have already been run. Mark's stance differs somewhat from Geoff's that it's a benefit to have pilot data for each experiment if possible. He says this because if you had to show pilot data for every experiment, there would be some experiments that you could not propose because you are unable to pilot them without grant funding.

When you are looking for a particular effect and the alternate outcome is an uninteresting null result, pilot data demonstrating the interesting outcome can really help your case. This is because you can essentially say that you've seen this effect in your pilot data so it's unlikely that you are not going to see the effect with additional subjects.

ALTERNATE OUTCOMES

Mark said that not having an alternate outcomes section is a fatal flaw for a grant. The purpose of describing alternate outcomes is to show reviewers that you have thought through the other possibilities and you aren't going to be dead in the water if your experiments don't work out the way you expect. Mark said the reviewers will always comment on a lack of alternate outcomes by throwing it back at you, basically asking, "What are you going to do if it doesn't work out the way you've planned?" and you will not have gotten a good score if that comment comes back. Even though it may sound trivial, bolstering your alternatives by saying that other groups have found a particular outcome contrary to the outcome you expect, especially if you are borrowing a task from another lab, is important to explicitly state.

FIGURES

I asked Mark how he recommends juggling the text space lost by inserting a figure and ensuring the figure is worth it. Although Mark likes Geoff's perspective that you can flip through the grants and look at just the figures, Mark agrees that that can be true for a manuscript, but he thinks that when you only have

12 pages for a grant, you can't tell the whole story in the figures. Mark says that if you just look at the figures in his grants, you'd be missing really important parts of the grant. Whenever something is more impressive in a figure, for example the timeline of the experiment, then he agrees that it should be put in a figure.

Mark uses the rule of thumb that a picture needs to enhance the text. So for example, if he writes, "faces are better remembered than scenes," there is no need for a figure to illustrate that point because it's efficiently communicated with words. In particular, Mark says that often the way data are presented in grants is a waste of space because authors often describe the data in words, and then include a bar graph, and then include a legend. Mark has trimmed his own grants by not including legends. He also thinks that now that everything is digital, and people are probably reading the grants on their laptops, you can probably make the figures a little smaller than you otherwise would have, because people can zoom in. He says if it's printed out, of course you need to worry about the size, but if you want to get some more space, the trick he uses is to shrink the figures.

Mark's very first MRI experiment had a very powerful figure and he's used that same figure repeatedly in grants. He thinks that figure has really helped to sell a number of his grants. Mark does believe that a figure can have that kind of power, so he reiterates Geoff's point of figures carrying importance, just as long as they are not redundant with the text.

SPECIFIC AIMS FIGURES

Mark has never had figures on the specific aims page. He feels really strongly about the specific aims page and has never thought of a figure that can capture the entirety of the grant. For example, if it was a curing cancer grant and you showed a scan with the tumor and without the tumor, maybe that would make sense, but you'd have to be thinking that globally.

WHAT IS OUT OF YOUR CONTROL?

Mark described the changing landscape in which the threshold for funded grants has gotten higher. It used to be that really great grants got funded the first time they were submitted. That basically never happens anymore. This means that having to revise your grant is not as serious or negative of an outcome as it used to be. Mark explained that if you are on a study section with 15 grants coming in for the second time, that means it's do-or-die time for those grants. They are extremely polished and have the advantage of having been revised based on reviewer comments. This means that your first submission may get punted down the line because of the very high quality of the ones ahead of you.

It is out of your control which other grants happen to be in the same study section as yours. Mark sees a lot of young investigators get panicked when their score was not high enough for funding. But he suggests that a lower score might not be due to the quality of the grant. It might just be due to your bad luck of being in a round where most of the grants are coming in for the second time, following revisions, so no matter how great your grant is, it'll probably be scored lower because the others are benefiting from reviewer comments.

WHAT IS IN YOUR CONTROL?

Mark said that something everyone should be doing is putting two institutes down when submitting an R01. For example, even if you are aiming for the National Institute of Mental Health (NIMH) you should also put down the National Eye Institute (NEI) or some other institute so that if it's in the fundable range and one institute doesn't want it, it can be moved over and considered by the other.

Don't submit your grant if it's not perfect. Mark finds that young investigators are often in a rush to submit their grants before they're ready. Before submitting, you should ask yourself *Is there really nothing more I can do to make this better? Is this truly the best I can possibly do?* He said that rarely is the answer *yes*, but it can be; it's possible to get to that point. You want to be at the point where you think, *I can't do anything else. This grant has got everything I have*, and that's when you submit it. It's really just a loss of four months if you wait to submit in the next cycle. In this scenario you only lose time, but if you submit a grant that is not ready, you might lose the opportunity to ever get it funded. This is because despite having the opportunity to resubmit based on reviewer feedback, a negative impression of the first submission may linger in the reviewers' minds when they consider your revised submission. In other words, rushing to submit a grant that isn't ready might actually be a flaw that is impossible to overcome.

Mark agreed that if you don't have a draft of your grant on paper one month before the submission deadline, there's no way you'll be ready in a month; there will not be enough time to get it ready by the due date. The last month before the deadline should be about making it as good as possible. It shouldn't be about writing the grant.

In terms of asking colleagues for feedback, he thinks that anybody who is willing to read it will give you good advice. Even colleagues who do not study your topic will help you know if it's in acceptable form and then the real experts can give you more nuanced comments. When submitting your first R01, you should give yourself plenty of time to get as much feedback as you need.

FIRST PERSON VERSUS THIRD PERSON AND VOICE OF THE NARRATIVE

When Mark provides feedback on his students' NRSA applications, which are training grants, he always crosses out *we* and puts *I* because it's not the lab, it's the applicant being funded by the grant, which is an essential focus of a training grant.

Mark agreed with Geoff that when writing an R01, you want to strive for simplicity of voice and he has become more informal as he's written more grants, frequently using *I* and *we* rather than *the PI* because he's acknowledging that some of the work should be credited to the entire lab, including postdocs and grad students. However, Mark is not shy about saying, "my lab has done this before," because he thinks it makes the grant more personal. Also, the grant is funding the person who has a track record so that track record matters and should be emphasized. However, Mark also reflected that the track record used to count for more than it does today. Mark said that it used to be the case that the track record had more of an impact than what was actually being proposed in the grant. For example, a well-known person's idea used to be relatively trusted because of their status in the field. Now, if the ideas aren't as solid as the younger investigators', then the more senior person's track record does not carry as much weight. Mark finds this a comforting change because he thinks it means the ideas are being funded more than the people, somewhat leveling the playing field for young investigators

On the other hand, if you have a track record with your R01, it's important to emphasize that. Mark said he just recently advised someone to make it clear that the ideas in their grant were an extension of what they were already doing. Mark said that when he reviews grants for a young investigator he wants to know that the line of work is really going to establish their research program and they're not just trying to get a grant for the sake of getting a grant. Mark views a grant more favorably if the ideas clearly evolve from what the investigator is already doing and is staking a claim for what they will do in the future. The alterative to this is someone putting together a grant on a topic because they know it's a hot method, which is not viewed as favorably.

WHAT TO DO WHILE THE GRANT IS UNDER REVIEW

Mark asserted that grants shouldn't get in your way and control your life. He says that the right thing to do while the grant is under review is to keep moving forward with your work, to the extent you can without the grant funding. Mark doesn't think it makes sense to play the game of sitting on data because a grant is under review. He said that the worst thing that could happen is that in a revision of your grant you say that you finished one of the proposed experiments, proving it's as good as you said it was, and now it's preliminary data for a future experiment, assuring nothing can go wrong in the new proposed experiments. Mark has seen revisions where people have a lot more preliminary data than initially, which means it's even better and strengthened relative to the first submission. In other words, moving forward with projects can only strengthen your grant down the road. Sitting on data or freezing the project to wait on the grant review process will slow down your publication rate, possibly jeopardizing more than just the grant.

FUTURE DIRECTIONS

Mark pointed out that an R01 is a 5-year grant. So even though the application involves proposing experiments that will be done in year five, by year five you're going in new directions. Proposing quality future directions is a useful, even critical, aspect of a strong application. Mark provided some examples of good versus bad future directions.

Good future directions would include listing a variety of additional analyses that will be possible with the data you propose to collect. For example, it can be productive to say that there is some theory or analysis that you can apply to the work down the line, which demonstrates that there will be value added through additional uses of the data. In other words, good future directions demonstrate the proposed experiments are bigger than the sum of their parts. Mark prefers future directions that build on the proposed experiments, showing where they're headed, outlining what you would do once you've collected the data in the proposed experiments, and briefly mentioning how the same data could serve multiple purposes.

Bad future directions include saying something that sounds important, but it's not something you can actually do. For example, you might suggest that you are going to use EEG, but you don't have a background in that technique or access to the equipment. Another problem would be mentioning an idea in future directions that is so good that makes reviewers wonder why you aren't proposing it in the main grant. Thus, future directions can be critical places to demonstrate your mastery of the literature and show you know the importance of your data. However, do not take it so far that they are detracting from the grant you are proposing.

HOW TO ESTABLISH YOURSELF AS AN EXPERT

Mark reiterated that your papers really speak for themselves in establishing the foundation on which your grant will be built. This is consistent with Geoff's comment that a grant is not the place to reinvent yourself. For example, Mark said that fMRI researchers establish new analyses all the time. By developing them in their papers they can say that, "The type of analysis that I created will be applied in this grant." Reviewers may not know about a new method or that you developed it. Your job is to make them aware of your impressive record.

Mark believes extra discoveries, such as novel analyses, are particularly important to mention because NIH R01s don't give funding for methods development. He suggests that every R01 include something along the lines of outlining how, as you conducted the funded work, you will be figuring out how to do this particular thing (e.g., advance computation) with fMRI. If you can add some methodological or analytical development that may come out of your research proposal, it is more value added and something that NIH will be able to get for free. In other words, if you develop a new analysis while working with your data collected from the proposed experiments, it's not what NIH directly funded, but it happened under their funding, so they will get acknowledged when it's published. This is a good way for a junior faculty member to keep funding coming with their publications after getting their initial grant.

SPECIFIC AIMS

On the specific aims page, Mark looks for a very accurate summary of what the grant will do: what the hypotheses are, how they will be tested, and what we are going to learn from it. He wants you to capture that in one page and then he essentially reads the grant to make sure that it's really true. Mark compared writing the specific aims page with writing an abstract for a manuscript. He pointed out that we all know that it's possible to write an abstract that is really bad and fails to capture what you actually did. Mark said an abstract can often read like you left it to the last minute and although it's perhaps inadvertent, it reads like you are lying about what the paper is really about. You want to avoid this with your specific aims page.

The specific aims page is something that you should not be writing at the end of your grant writing process and definitely not something you spend the least amount of time on. Rather, it should be something you take very seriously and spend a lot of time on. It is the first thing reviewers read and it's either going to set them up positively or negatively. Although it might be possible to start with a negative impression and be swayed by the grant, it's certainly preferable to start with a more positive impression and then look forward to the detail of how you will accomplish these really fascinating aims. Mark has done an interesting exercise while reviewing grants where he ranked the grants after only reading the specific aims page. Then he read the grants and discovered that his initial rankings based on the specific aims page ended up being very similar to how he ultimately ranked the proposals after reading them in their entirety. This outcome suggests you should write the specific aims page like it is the most important part of the proposal someone will read.

Mark added that there are now foundations that fund people based on one-page proposals. Mark's conversations with the nonacademics who evaluate these proposals basically say that if you can't sell your idea in one page, you're not going to be able to convince them with 12 more pages, so why not have a grant that's only one page? From that point of view, the specific aims page is the page with which you've got to sell your reviewers on your grant, and quite frankly, that makes a lot of sense.

Mark suggested writing the specific aims first and once they're clear, you can move on to the proposal. Of course, when you are done with the proposal, you should go back and make sure the specific aims page is consistent with the proposal you wrote. Another advantage to starting with the specific aims page is that you've essentially just written the grant in your head and then you only have to flesh it out on paper and write 12 more pages. He thinks this is a great exercise for everyone, especially junior faculty who might be daunted by starting on page one of twelve and feel stressed about coming up with 12 pages of new material. When you have the specific aims page written, expanding it to 12 pages doesn't feel like such a daunting task.

HOW TO BE REPETITIVE IN A USEFUL WAY

The specific aims page is the first time you tell the reader something. You then have two more times to tell them. You tell them for the second time at the beginning of the background. Often the first couple of lines of the background are the same as in the specific aims, so now the reviewers have seen it twice. And then the third time is when you get into discussing the analyses and predicted outcomes. You aren't literally repeating outcomes, but in a subtle way you are because you're going over what you predict the outcomes will be, repeating yourself in case the reviewers forgot what they were.

Mark said that sometimes he writes a grant where he feels the need to summarize a whole specific aim. In this case, at the end of the specific aim, he provides a summary to make sure the reviewers are following him. For example, he explicitly says, "This specific aim is trying to determine whether XYZ is processed in the frontal cortex." Then, he segues into the next aim by saying that

"The previous aim was about the frontal cortex and this aim is about the striatum, so together we are going to cover the frontal cortex and the striatum." Mark said you want to think about the fact that reviewers just read through 5–6 dense pages, and it's helpful to have these moments where you step back and give them a summary. He acknowledges that when you only have 12 pages, sometimes it's hard to find space to give short summaries like this, but if you can find a place to help reviewers put it all together, it can be really helpful.

Along the same lines, Mark likes to have very specific headings. If the reviewer takes a break, which they always do, and then comes back to keep reading your proposal, your aesthetics are set up to deal with the fatigue that reviewers experience when reading many proposals.

SOLICITING A CO-PI

Mark said that you should only solicit a co-PI if it makes sense for the science. But he really encouraged doing so under certain circumstances. He thinks that if you are at an institution where there is another great person and together you can write a really strong grant, then you should definitely do it. For example, if you know A and they know B and when you put them together you could have something really great (like chocolate and peanut butter), something even better together than they are individually, it just makes sense to write that grant. Mark is concerned that junior faculty may choose not to seek out these obvious, generative collaborations because their department chair tells them it would be bad for their tenure case. He thinks that's folklore and if there is a natural collaboration you should go for it. It may or may not be your primary R01, but good science just makes sense.

Mark has collaborated on grants with his trainees when they want to propose something new but they are moving to a new institution and don't have their equipment yet. They will propose to do it in Mark's lab until their lab is set up. Mark's hope is that rather than hurt their shot at tenure, it will actually help them get tenure because it's helping them to do something they would not otherwise be able to do, or it combines conceptual ideas that are better than theirs alone.

Mark said he wants to make sure junior faculty members do not bring on a co-PI just because they aren't getting an R01 and want to add someone who will help them get it. The idea is that if a co-PI makes sense for the science, then do it. Don't listen to people who tell you that it is bad for your career. One reason that there might be a stigma among more senior faculty about writing grants with a co-PI, especially one at another institution, is that there used to be more concern about how collaborations could be possible across long distances. Years ago, in one of Mark's first grants, he proposed a collaboration between himself at Berkeley and someone just across the bridge, in San Francisco. The negative feedback they received was basically that the reviewers couldn't imagine how the co-PIs were ever going to get together. Now that data can be placed in the cloud and everything is electronic, you are not likely to be criticized based on the geographical distance of your co-PI, assuming you have a good working relationship and can demonstrate that to the reviewers.

WHAT TO DO WITH A REVISION

Mark emphasized that one possibility is that reviewers will like your grant if you respond properly to their concerns. But he also pointed out that it is possible that they are just never going to like it. Perhaps it was simply a poor fit with those reviewers. If it's really clear that your grant wasn't a good fit with the reviewers, Mark wants you to be aware that the NIH has the option of going to a new study section as a new submission without your grant having any history. Mark hasn't done this himself but thinks that it sounds like a reasonable approach. Generally speaking, he thinks that if you aren't in a rush, you might as well try it again with the same reviewers. However, if it's clearly dead in the water and there is no way to sway the reviewers, submitting it as new is an option.

SERVING ON A STUDY SECTION

Mark said you won't ever really know what it's like for your grant to be reviewed unless you are sitting in the room when people are doing it. He thinks everyone should serve on a study section at least once in their career. Unfortunately, it's a thankless job with poor incentives for reviewers. The only reason people do it now is probably for the experience because it's a ton of work—reading 10–12 grants, three times a year. If you do it your first year as a faculty member, when you are creating your first course, that would be an overwhelming burden.

Once your courses are prepped and you feel like you can make the time sacrifice, there are real benefits to serving on study section. One way to manage the experience is to volunteer to be an ad hoc reviewer, where you volunteer to do it during one of the cycles. Then, you can make the decision later if you want to be a standing reviewer. Mark said they are always looking for reviewers so it's unlikely they will turn you down if you volunteer for just one session.

DIFFERENCES BETWEEN NSF AND NIH

Mark said that you should be aware that NSF and NIH grants have different styles and different priorities. His advice is that whatever idea you have for your R01, you should also try to see if NSF would be interested in it, either in its entirety or even just some piece of it, because there is nothing stopping you from submitting to both institutions. You might find that you only need to modify it slightly to submit to both institutions. It is generally true that NSF doesn't have enough money for large labs that are doing expensive imaging or have big operations, but it's a possible source of support and it all just depends on the topic.

At NIH, you need to sell why your research is relevant for the health disorders that the specific institute seeks to understand and treat. You can't just offhandedly say that there isn't anything clinically relevant in your grant because it's about basic science. The clinical relevance of your basic research is typically as simple as gaining a better understanding of the disorders. For example, patients with schizophrenia suffer from a host of cognitive deficits that can be better understood by studying how mechanisms like attention and working memory operate normally. Obsessive-compulsive disorder involves continually performing the same task, meaning that understanding mechanisms of cognitive control might be particularly relevant. Frequently, a few days spent with the background literature on the disorders handled by a particular institute is sufficient to have a strong section in your grant on clinical relevance.

CREDITS

Fig 10.1: Adapted from Robert M. G. Reinhart, Julia Zhu, Sohee Park, and Geoffrey F. Woodman, "Synchronizing Theta Oscillations with Direct-Current Stimulation Strengthens Adaptive Control in the Human Brain," *PNAS*, vol. 112, no. 30. Copyright © 2015 by National Academy of Sciences. Reprinted with permission.

CHAPTER 11

How to Run a Lab Without Graduate Students

by Ashleigh

There are many times throughout your career when you might be faced with trying to run a lab without the assistance of graduate students. Here are just a few examples. You might be hired at a school in the psychiatry department on soft money and all your funds are spent on your full-time RA. You might be at a liberal arts school that gives you high priority for a graduate student but you can only make one offer each year and the person you offered to take might not take the offer. You might have run a wildly successful lab at a R1 institution, but you have been so successful that you have a year or two when all your graduate students and postdocs get jobs, leaving the lab empty. You might have a visiting position that does not come with graduate student trainees or funds. You might work at a small liberal arts school where there is no graduate program. The point is that most of us have been faced with running a lab without graduate students at some time. Indeed, all the scenarios described above are ones that I have been faced with or someone I know has found themselves in. Below I describe a number of ways in which I have personally dealt with these situations. In my case, I have gone an entire career without having dedicated graduate students supporting my research, so I hope to provide useful advice about keeping research programs going with alternative personnel. I have always run my lab with only undergraduate students so the advice here will primarily focus on overseeing undergrads in the lab and collaborating with graduate students in other labs.

I begin with some strategies to form partnerships with entities that do have the resources some types of graduate programs have and then turn to a set of solutions that are within your control. These are suggestions about how to tailor your local environment to allow you to succeed in the absence of working with people earning their PhD. I want to add the caveat that some committees for promotion and tenure may view your productivity less favorably without a graduate student. For example, deans can frown upon your weak mentorship record if you have not graduated any PhD students. A track record of mentorship is important for getting grants as well. The advice in this chapter is for people who don't have graduate students due to circumstance rather than choice. If you have the option of training graduate students and are working toward tenure, it might be wise to consult your committee or chair about whether your accomplishments will be viewed less favorably because you did not select a graduate student when given the chance.

GIVE A TALK IN THE PSYCHOLOGY DEPARTMENT

One of the reasons you might be running a lab without graduate students is that your appointment does not come with funding for graduate students. If this is the case, I suggest asking to give a brown bag talk in your department of choice, like the psychology department. This is essentially advertising your work and letting the students and faculty know you are there. Doing so might attract faculty collaborators or graduate students who work with another faculty member but are looking for ideas or a possible change of course in their own research career. If nothing else, it will likely generate some great feedback and interesting questions about your work.

CREATE A PARTNERSHIP WITH A LOCAL SLAC

As a former faculty member at a small liberal arts school, I know that often there are such heavy demands placed on teaching at these institutions that faculty can struggle to offer high-quality research experiences for their undergraduates. However, the importance of these experiences is undeniable when it comes to graduate school admission. If you are in a position where you do not work with undergraduates, for example at a hospital, and you want some undergrads to work for you, one option is to contact local small liberal arts schools and offer research experiences for a select few of their best undergraduates. You will likely get an excellent response. This would benefit their students in many ways, including that it would allow their students to get a letter of recommendation from someone at another institution. You could also potentially appeal to the administration at that institution, such as the academic dean. Explain how research experience is a vital part of getting into graduate programs and how this opportunity would ultimately be beneficial to their graduation statistics that track how many graduates get into graduate school and subsequent employment.

CONTACT MASTER'S PROGRAMS THAT ADVERTISE RESEARCH EXPERIENCE

You may already realize that some master's programs in psychology position themselves as steppingstones into PhD programs. In reality, some of these programs are very popular and become sufficiently overwhelmed with students that they are unable to offer all their students the high-quality research experiences they were promised, and that the students need to ultimately get into their desired PhD programs. You might be able to position yourself to help them and help yourself by offering to work with some of the students. You can contact master's programs that advertise themselves as stepping stones and explain that they would benefit from you offering experiment ideas, mentorship, authorship on conference presentations, and so on, if their students could participate in research in your lab and collect some data for you. You could even do this long distance provided they have the equipment needed to collect data. There is no reason that these programs need to be local as long as you have confidence in the faculty mentor who oversees the students. I have worked with a student in a master's program in a different state and the collaboration worked out very nicely. I also got to know the head of the graduate program overseeing this student and was invited to give a talk. It was a great networking opportunity for me, in addition to the other advantages mentioned above.

APPLY FOR SUMMER RESEARCH PROGRAM GRANTS

There are plenty of grants that fund summer research; for example, the APA and Psi Chi both have summer research grants, where students can come from all over the country to work with you for the summer. These don't have to be students from your home institution, which is a benefit if your home institution is non-traditional or does not have an ideal student body from which to draw research assistants. One idea is to advertise at other institutions for someone to apply for a Psi Chi summer research grant[1] to come and work with you. Alternatively, to find such a student you could advertise at your home institution in upper-level classes, speak at the Psi Chi meetings, reach out to your friends who are at small liberal arts schools and see if any of their best students want to apply, or even contact your alma mater and inquire about the summer plans and research interests of their top students.

NETWORK AT CONFERENCES

Given that I have never had dedicated graduate students, I have always had more ideas than I could possibly pursue with my limited resources. It is often easy to lose sight of the fact that everyone is not in the same boat as me when it comes to having more ideas than labor. Don't forget that there are some people who have more students or postdocs than they have ideas and if you have an excellent idea or are putting together a promising grant, other faculty might be happy to work with you to gather pilot data in exchange for some monetary reward or support, pending grant funding. Engage in as many conversations as you can at conferences, with people over dinner, at coffee hour, in front of posters, and so on, to try and locate collaborators. I've been surprised where some of my opportunities have come from; go to conferences and be social!

---

1   www.psichi.org/page/summerresinfo

CREATE A CONTRACT WITH COLLABORATORS

In order to embark upon a mutually satisfying and productive relationship with collaborators who are not your graduate students, it makes sense to have a contract outlining expectations. This is probably too formal and unnecessary if you are collaborating with other faculty colleagues or peers, but if you are working with undergraduates especially, it's important for everyone to be on the same page. Your institution might have an undergraduate research office that offers a suggested contract. A contract specifies many issues that are wise to make explicit and can cause problems when dealt with implicitly. Specifically, the contracts state who will mentor the student (if not the faculty member), who (and when) the student should contact for questions and concerns; how frequently the student should be attending lab meetings and other events, such as departmental brown bags; how many hours per week the student needs to put into the research; how the student is compensated (e.g., paid hourly, awarded credit toward graduation, strictly volunteer); how their grade is determined if the student is being graded on their work in the lab; the training required to work in the lab (e.g., human subjects certification); expectations of professionalism when working in and representing the lab; the skillset the student is supposed to develop over the research experience (e.g., programming, data analysis); anticipated accomplishments, such as writing a grant, submitting to a conference, or co-authoring a submitted manuscript; and the specifics of the assigned research project (e.g., title, methods, planned analyses, author order, if applicable). Formalizing these expectations, goals, and requirements in a document that the student and the faculty mentor both sign can avoid many headaches down the road.

PREP DURING SUMMER OR WINTER BREAKS

You do not want every semester to start out with your undergraduates waiting for experiments to be ready before they start collecting data. Get this prep work done ahead of time. Plan ahead during the summer months or winter break so that all the prep work doesn't eat into the first few weeks of the semester and slow down data collection. These breaks are a great time to have your students send their schedules and figure out a common time that everyone can attend lab meeting. Before the semester begins is also a great time for students to complete human subjects training or conflict of interest disclosure forms if they need to be certified to run subjects in the lab. I spend time in between semesters getting pictures of their student ID cards that I need to submit to get them swipe access to unlock the lab doors. I want to start data collection as soon as possible when the semester begins, so I try to plan ahead to overcome any hurdles that I can anticipate over break.

FILL YOUR LAB WITH ALL THE OTHER RANKS

When Geoff experienced dry spells because his graduate students got jobs and left, he has filled his lab with people from other ranks. He has employed paid research assistants, welcomed more undergraduates into the lab than usual, continued his collaborations across the department, and recruited postdocs. Graduate students can be great, but they simply are not the only way to get work done.

There are faculty positions in medical schools and other institutes that may not allow you access to a graduate student pool based on your appointment level. Given that these jobs typically require you to pay approximately half of your own salary from grants, this is not such a big deal. That is, you will need to have a grant to survive in these jobs anyway and with that grant money you can hire full-time RAs and postdocs who can program experiments, collect and analyze data, and write the first drafts of manuscripts. As you will see in the interview at the end of this chapter, this can be a very productive personnel model.

Typically, when you cannot get a graduate student, you imagine that working with one is better than it actually is. As Geoff mentioned, approximately half of all graduate students in doctoral programs across the United States leave graduate school without getting a degree. Among the half that do finish, a large percentage will leave graduate school with a very small number of publications from their 5–6 years in the program (i.e., one to two publications). Given the amount of time that you dedicate to these students as a mentor, even the best graduate mentor can sometimes question whether the expected payoff justifies the work that goes into mentoring graduate students. Usually, when we think we need a graduate student in the lab, we are thinking of that once-in-a-lifetime student who publishes 15 papers in glossy magazines and changes the climate in the lab. Given how rare that is, the strategy of staffing your lab with RAs and postdocs is much less risky. Although statistics are lacking, the common impression is that RAs and postdocs have a higher success rate than graduate students. A whole book could be written about why this happens.

LAB MEMORY

A common way to best utilize undergraduate RAs is to require that they work in the lab for more than one semester. I know some colleagues require students make 3-4 semester commitments to the lab. The general idea behind this approach is to prevent a high turnover rate each semester, which can require a significant amount of your time to train new undergrads, get them added to the IRB protocol following human subjects certification, and so on. Although I don't make this requirement because there really is no way to hold students to such a commitment, I do prefer to work with sophomores or juniors rather than seniors since I know the senior won't be with me longer than a year.

My friend Molly Erickson employs a few methods to build what she calls *lab memory*. She has created a lab manual with thorough documentation about experimental protocols, as well as how to do various lab tasks like replenish subject money, reserve rooms, and make copies. Molly created MATLAB tutorials and a repository of important journal articles to get new RAs up to speed on background literature and programming skills needed to be productive members of the lab. Veteran RAs can then assist new lab members with these skills independently from Molly, which frees up her time from repeating these intensive training tasks each semester.

I have done something similar, using veteran RAs to train new RAs. I have also found that small tasks like ensuring there are enough copies of informed consent documents can be problematic if they run out while I'm at a conference or given no warning that we are low, so these tasks are really important to delegate to other folks in the lab. Finally, I once actually caught errors being made by a previous RA when I asked them to revise and update the protocol for an experiment they were running. Having assistants write and revise protocols that I can double check is now my method for ensuring that lab memory is built, and there is some consistency in the way everyone is trained to perform their tasks. These suggestions ultimately mean that you should determine what tasks are the most time consuming for you and figure out if there is a way to build them into lab memory.

RUNNING A LAB AT A LIBERAL ARTS SCHOOL

One common reason faculty end up running labs without graduate students is because they work at small liberal arts schools. As I mentioned an increasingly common way to measure a department's success at SLACs is to ask about the proportion of students accepted into graduate school. Given that research experience is related to admittance into graduate programs, providing quality research experiences is integral to this metric of departmental success.

Relative to the teaching load carried by faculty at R1 institutions, it can feel nearly impossible to conduct research while teaching multiple courses, advising, and completing committee work often required by liberal arts institutions. One interesting way to continue the momentum of your research program is to incorporate it into courses as a research project or extra credit (e.g., for participation) or even create an entirely new course that involves working in your lab.

When I worked at a small liberal arts college, the department revised our statistics and research sequence of courses to help our students get into graduate school. One of the changes we made to the curriculum that was mutually beneficial for both the faculty and undergraduate students was to offer a third course in the sequence that was an advanced research course. This course was essentially working in our labs, giving the students graduate-level experience and responsibilities. The main benefit to the faculty members was that this four credit-hour course was part of our 12 credit-hour teaching load for the semester and we spent it overseeing and conducting research in our respective labs, rather than having to conduct research on top of the heavy teaching load. The students received a number of benefits from this reorganization of the course sequence as well, such as working on generative, publishable research projects, presenting at conferences, co-authoring publications, getting to know a faculty member outside the classroom for letters of recommendation, and being able to report in graduate school interviews that they were essentially functioning as a RA or graduate student in a lab.

There are many tasks you can train an undergraduate student to do in the lab through a class. I once created a course called Computer Programming for Psychology Majors. This class taught students to program using E-Prime, the software that I used for stimulus presentation in my lab. In this scenario I was able get experiments programmed by teaching students to do it as part of my teaching load! Another course-related way to have undergraduate students assist with your research is to teach a course on scientific writing and have them work with you as you prepare a paper or grant you plan to submit. Or you could teach a course on research methods and have them either run subjects for your experiments or have them participate in your experiments as subjects for extra credit (provided you have IRB approval to do so). All these activities benefit you and the students.

Let's return to the concern of not having enough time to conduct research due to your teaching load. You really do have time to conduct research and juggle a 12-hour per semester teaching load. I did this myself at a number of different institutions while still juggling a family of three young kids. It means that you might spend less time on lecture prep than your peers, and you might lean on grading assistants more than your peers, but you can make the time for research if it matters to you.

THE BENEFIT OF RUNNING A LAB WITH UNDERGRADS RATHER THAN GRAD STUDENTS

The concept of running your lab with only undergraduates and no graduate students is a reality at most teaching institutions. I argue that you can be quite productive in your lab, even without graduate students. Although not having graduate students certainly means that a bulk of the programming and manuscript writing falls in your lap (that same lap that has a much higher teaching load than your colleagues at research-based institutions), there are some distinct advantages to working with undergraduate students over graduate students. The purpose of this section is not to argue against how wonderful it can be to have a fantastic graduate student or the many benefits of working with postdocs. I sympathize with the many demands on your time at a teaching institution and recognize the many frustrations of trying to maintain a productive research program, and even more so maintaining a research program that is competitive with your peers at research-focused institutions. However, Geoff and I have had many conversations with each other and with colleagues that boil down to three major advantages of running a lab with only undergraduate students.

First, undergraduates have a much shorter time frame during which they need to succeed in order to impress you to get that letter of recommendation they desire. They have real pressure to gain admittance to a graduate program in the next year or so. That means that, unlike graduate students, they are very likely to hit the ground running. The urgency of undergraduate students to impress you with their work ethic and productivity contrasts with that of many graduate students who have just gotten into graduate school, essentially just achieving their biggest goal, and who may settle in thinking that they have arrived, giving themselves permission to slow their rate of work.

Second, your commitment to work with an undergraduate is usually renewable each semester. Unlike the five-year commitment you make to PhD students, you can easily stop working with undergraduates after just one semester if their role in the lab is not mutually beneficial. By working with undergraduate students for a shorter period of time, you spend far less time going through the ups and downs of life with this person, and hence can just capitalize on productive periods with students when they are really able to focus on your research. This emphasis on the benefits of shorter working relationships with undergraduates sounds heartless, especially to me because I was very grateful to have a wonderfully supportive and understanding PhD advisor as I dealt with the death of both my parents and the birth of two children during my PhD program. However, it is true that such life events impaired my productivity over time. The shorter time commitment of working with undergraduate students can lead to many short but productive working relationships rather than the fewer, longer relationships most faculty have with graduate students that are understandably interrupted by life happening.

Third, not having a graduate student can increase the efficiency and oversight of the lab by cutting out the middle person (i.e., the graduate student who is usually overseeing the undergraduate students collecting data). If you as the faculty member are in regular contact with the undergraduates who do most of the actual labor, you have a much greater degree of oversight and control over issues such as troubleshooting any technical problems, promoting a positive lab environment, and verifying that, for example, lab equipment is properly maintained, supplies are ordered when necessary, and copies are made when informed consent forms run low.

You might be thinking that some of these benefits of working with undergraduate students are time consuming and therefore, do not feel like real benefits. On the front end, it may seem as though managing all the undergraduate students working in your lab by yourself, or continually training new undergrads whose time commitment is shorter than that of graduate students, wastes more of your time than having graduate students would. I agree that not having a graduate student oversee these operations does seem more time consuming on top of your already heavy teaching load. However, I argue that the degree of issues that can arise by losing oversight of the regular operations in the lab or the issues that come up when working with graduate students over a long period of time can have similar, if not significantly greater, impact on lowering your research program's productivity.

WHY YOU SHOULD BOTHER

So why should making time for research matter to you if you have a faculty position in which teaching is your primary responsibility? Like we said earlier in this book, you might not start at your dream school. Or the school you originally love might not work out for a variety of reasons. For example, an administration change could create a climate that makes the school a less ideal place to work, a change in your personal life could require a geographical change, or you could simply outgrow your position, becoming competitive for jobs at more research-focused institutions. The truth is that we do not know what the future holds, and situations can change due to our success or issues outside our control. By continuing to publish, you stay mobile and competitive for any of life's curve balls. In addition to staying marketable by conducting research, you might get a teaching load reduction if you are invested in research. Similarly, you might get out of undesirable service positions because the dean or chair sees the value in your research productivity and wants to protect your time.

Maintaining scholarly activity in your field also benefits your students. For example, if you remain visible in your field by publishing and attending conferences, the letters of recommendation that you write for your students who are trying to get into graduate school will carry more weight. Indeed, attending conferences and publishing promotes your institution and thereby your students, in addition to yourself.

Ashleigh's interview with

search.bwh.harvard.edu/new/index.html

Jeremy Wolfe, PhD, Professor of Ophthalmology and Radiology at Harvard Medical School, is the Principal Investigator of the Visual Attention Lab, an affiliate of Harvard Medical School and Brigham and Women's Hospital. We wanted to interview Jeremy because he has run a wildly successful lab outside of the traditional tenure-track R1 structure, without graduate students. Jeremy is known for having a very large number of successful people come out of his lab and we were very curious to find out what was unique about his approach to running a lab.

THREE LEVELS OF PERSONNEL

The personnel in Jeremy's lab are from three main categories. There are paid RAs, traditional postdocs, and visitors who range from high school students to faculty spending their sabbatical in Boston. The first two categories, the RAs and postdocs, make up the core of his lab and are more feasible for early career faculty, so I am primarily going to cover Jeremy's input about those two categories here.

FULL-TIME PAID RESEARCH ASSISTANTS

The similarity between Jeremy's lab and a standard R1 lab is that his RAs function as junior graduate students. The RA positions are filled through a call for a research assistant he sends out to about 100 of his colleagues and posts on the internet. That posting for an RA usually receives about 70 applications each year. Jeremy thinks that the healthy number of applications he receives for this position is probably due to a combination of factors, including that Boston is a prime location for young adults who are looking for a temporary job while they decide what to do after their undergrad years, and that his lab also has a good reputation for helping individuals in these RA positions get into grad school. Because Jeremy knows that graduate school applicants are much more likely to get into a good program if they spend a couple years working in a lab rather than applying straight out of college, he takes grooming his RAs to be well positioned for graduate school seriously. To this end, he asks for a two-year commitment with the idea that in the second year, he hopes and expects that they will be working on their own projects.

I asked Jeremy how much freedom he gives RAs on their projects. He clarified that, while he would like them to come up with a project of their own, since these positions are grant funded the lab has money to conduct only certain types of projects. That means that in his RA's second year, the goal is to have an RA who thinks of the next clever experiment to do in a series of grant-funded experiments that is ongoing in the lab. When this successfully occurs, the RA can take the lead and serve as first author; it works out well for the RA and for the success and momentum of the research program.

Jeremy's lab is known for having a large and accomplished group of alumni in the field. I asked Jeremy what he thinks contributes to the successful careers of many people who were originally RAs in his lab; is it the selection criteria he uses to hire RAs or the mentorship and training they receive once they are in the lab. He tends to think it's both: that he brings in good people and they are well trained. Jeremy reflected that having a healthy number of successful RAs may actually be a consequence of not having graduate students in his lab because some of the attention that would have gone to graduate students instead goes to his RAs. This means not only Jeremy's attention, but also attention from relationships with others in the lab, such as postdocs. Thus, RAs in his lab enjoy time and attention that would go to graduate students in a typical R1 lab situated in a department with a graduate program.

Jeremy also seems very accepting of the fact that not all of his RAs go off to graduate school, although a good percentage of them do. As we've discussed elsewhere in this book, he reiterated that a good reason for being a RA is to figure out if you like the academic life before going to graduate school. He pointed out that there is no shame in going from being a RA to something else, such as culinary school, whereas if you go into a career in a kitchen after being in graduate school, you might feel like a grad school dropout. Jeremy's assertion that he knows he is running a finishing school for eventual grad students, combined with his relaxed acceptance that not everyone is going to decide they want to enter graduate school, gave me the impression that this moderate approach to managing his lab must contribute to its success.

POSTDOCS

Jeremy said that his postdocs are usually people who approached him and told him they'd like to be a postdoc in his lab, rather than Jeremy doing much advertising or recruiting for these positions. When approached in this way, he often responds (when he can) by saying that he can offer funding for the position for one year, and suggests that after that year, they write an NIH fellowship grant or find some other funding mechanism to fund additional years of support. He has found a good amount of success with this approach over the years. Being approached by graduate students who want to serve as postdocs in your lab is something that is likely to happen only to more seasoned faculty members with an established reputation of being a positive mentor. This ease of staffing the lab with qualified individuals is one of the reasons I was interested in Jeremy's approach to lab management.

VISITING GUESTS

Jeremy's lab often has several varieties of visiting guests. One type is visiting graduate students (e.g., the Chinese graduate student system has a mechanism for sending students overseas for a period of time). Another type is the persistent and interesting local high school student who is looking for research experience. There are also faculty who are interested in spending their sabbatical in Boston and working with Jeremy. A final type is the individual from one of the

European systems that have a model of doing a master's before a PhD. Some of these programs require that some of the student's master's training is spent in someone else's lab, even overseas (e.g., Jeremy has had quite a few students from Munich under this mechanism). Typically, these individuals essentially invite themselves to visit Jeremy's lab and have their own funding. I asked Jeremy about his decision process regarding whom to allow into his lab when they invite themselves under this fourth category. He said that his decision process is not particularly formal. Rather, he talks to the candidate and decides if they seem like a good fit. He pointed out that because he isn't horribly crunched for space, the worst thing that could happen under these circumstances is that nothing productive comes out of time with this visitor.

I asked Jeremy what he would suggest researchers who are not as established and don't have people approaching them to be postdocs and visitors. His response was that he would go out and get himself some money! Jeremy confirms that money cures a lot of ills in this regard, because then you've got the potential to recruit a good postdoc.

WORKING WITH UNDERGRADS

Jeremy said that he likes my emphasis in this chapter where I suggest that faculty can go a long way working with undergraduates in their lab. One difficulty he sees with undergraduate RAs is that they also have to pass their courses and they are not full-time employees. However, he notes, a good senior is just 12 months younger than the applicant he will hire as a full-time paid RA. For me, this point really emphasizes that a main functional distinction between an undergraduate working in a lab and the full-time paid RAs working in Jeremy's lab is that the undergraduate working in a lab has other demands on their time.

Jeremy also mentioned that the luckiest undergrads find themselves in labs where they are encouraged to submit abstracts to major conferences in the field and are maybe even financially supported to attend those meetings. He said that many of the successful applicants for his RA positions can say they went to the Vision Science Society or Society for Neuroscience meeting and tell him something interesting about their submission.

SUCCESSFUL RA APPLICANTS

I asked Jeremy what causes an application to rise to the top of the pile of the 70 applications he gets for his RA positions, besides experience at such international meetings. Jeremy said that first, it helps if the applicant has done some research as an undergrad and can say something intelligent about the research they've conducted. He said that second, the recommenders matter. He asks applicants for email recommendations from two to three people who can speak well of an applicant. He pointed out that it's a bad idea if the person asked to write the recommendation is someone like a soccer coach. A soccer coach may be a wonderful person and a shaping influence on someone's life, but, for this application, his or her letter is just not going to be that useful. It is better if you have a faculty advisor or two who can explain why an applicant might be a good fit for the lab. Implicitly, this means that he's essentially partially judging the applicant's fitness by the quality of the recommender. Jeremy said that third, CVs matter. He advised that the resume applicants put together for a job at the bank is not the same as the CV applicants want to use to apply to a lab. He said that many students seem to go to the career center at their university, where they are advised to start their resume with a career objective. That is probably good in some situations, but it sounds a bit odd when a resume says something like *Career Objective: to get a job in a lab that studies visual search*. Another example of where applicants can go wrong with their CV is listing outdated accomplishments. That sixth-grade spelling bee championship was great, but it can be retired from the CV when applying for an RA position.

When he sorts through RA applications, Jeremy pays a lot of attention to the input of his current RAs. He shares the applications with his current RAs and when they sit down and discuss who is on their short list, their short lists are almost always very similar to his. A high correlation between his short list and theirs means that Jeremy is not just picking people based on, for example, knowing the faculty who the applicant worked with as an undergrad, but there are multiple factors that everyone seems to weigh similarly. I was so happy to hear this because I always want to believe that students who attend liberal arts schools with less famous faculty still have a shot at these positions.

I asked Jeremy if applicants must know how to program in order to be considered for RA positions in his lab. He said that it is a plus, but it is not a requirement. Indeed, sometimes it turns out that a bright young high school student there for the summer can program circles around everyone else, so rather than require programming skills for his RAs, Jeremy likes to use the talents that people have. He said some RAs come as great programmers and he will use them as programmers. If an RA doesn't know how to program, he encourages them to learn, in part, to make them competitive for grad school. Certainly, Jeremy wants his RAs to have some sort of skill that will be useful in the lab, but he is not super picky about what the skill might be. For example, a talent for interacting with difficult subjects might be just as useful to the lab as programming skills. This is one reason that the current RAs are so useful in the selection process.

They know more about the job than Jeremy does since they are the ones doing it every day. Jeremy said that when he has hunted for postdocs, he would similarly get other postdocs in the lab at the time to interview the person because they are currently doing the postdoc job and will have a good idea as to whether the applicant would be a good fit for the lab.

BENEFITS TO HIS SYSTEM

When I asked Jeremy for some of the benefits of running a lab without graduate students, he responded that graduate students, perhaps more than postdocs, are in some ways your responsibility forever. He said graduate school mentors typically have the primary responsibility for nurturing a grad student's career and that can be hard if they aren't fantastic scientists. This benefit of not having grad students is consistent with the lower degree of commitment made to nongraduate students (i.e., undergrads) that I mentioned earlier in the chapter.

Jeremy was also quick to point out, however, that you have a responsibility to anyone who is in your lab, from high school students on up. He pointed out that labs that get a bad reputation are the ones where PIs take advantage of people working in their lab. In such labs, it may be surprisingly difficult for graduate students to ever get their degree or for postdocs to ever get a job. Jeremy reported that he often feels a bit bereft when a good postdoc leaves, but he realizes that his job is to make it possible for them to leave successfully, at the right time. Even if it might not be the same as the relationship of graduate student to advisor, Jeremy still says the people in his lab become a sort of family and he wants the family to thrive, even when they leave home. Given that there are some people better suited to that role than others, Jeremy's ability to mentor and work well with all the people in his lab might also explain some of Jeremy's success as a mentor and PI.

UNIQUE SUMMER PROGRAMS AND HIGH SCHOOL STUDENTS

Jeremy has a substantial collection of high school or college kids working in his lab over the summer from a variety of programs (e.g., the Research Science Institute based at MIT is a ferociously competitive summer program with the goal of giving research opportunities to "80 of the world's most accomplished high school students" according to their website,[1] or Project Success which is directed specifically at giving biomedical experience to kids from the Boston area who

---

[1] https://www.cee.org/research-science-institute

might otherwise not have such experience[2]). Jeremy said that despite these two programs pulling from vastly different groups, summers with combinations of students from these programs and elsewhere typically go very well. The postdocs and RAs are the primary mentors for these students over the summer. This becomes mentorship experience they can talk about when they apply to graduate school or are on the job market. In other words, Jeremy said that the summer experiences are good for the students visiting the lab and good for the existing members of the lab. I asked Jeremy if that's why he participated in these programs or if he gets any science out of these summer kids. He said that for summer programs, you really have to come up with a project that will generate data in basically a month so there is something for the student to write up at the end. This may sound impossible, but in visual psychophysics, which Jeremey's lab does, it can certainly be done. Indeed, he has a slew of high school student authors on papers who made significant contributions to the work.

Working with student authors raises the issue of authorship. Jeremy said that he tries hard to keep track of projects as they go along in terms of who makes what kind of contribution. The basic rule he applies in the lab is that simply running the subjects doesn't make you an author; rather, you need to make some variety of real intellectual contribution, consistent with what we've discussed elsewhere in this book. He conceded that looser criteria may be applied to posters than to papers. Jeremy said that ultimately, talking about authorship early and often is important so that there aren't surprises. He thinks it's important not to give away authorships because you're too lazy to think about who deserves it, but you also don't want some relatively shy student not to end up as an author because the student didn't think they could mention to the PI that they really had done all the work on a particular project.

OVERALL IMPRESSION

During our interview, I kept wondering who Jeremy was referring to when he'd answer one of my questions using the pronoun *we*. At the end of the interview, I suggested to Jeremy that one of the reasons his lab was successful without graduate students was that he clearly takes a collaborative approach with his lab members, as evidence by his answering all my questions by saying *we* rather than *I*. He agreed that the lab operates as a committee of the whole where nobody has to feel like they are allowed to work on only one particular project. He said he encourages people to be as collaborative as possible rather than end up in silos. Jeremy said that perhaps he runs the lab in this way because he came

---

2 https://mfdp.med.harvard.edu/dicp-programs/k-12/high-school-programs/project-success

from a similar lab, one that had the feeling of a collaboration. It was clear that his advisor was in charge, but not in an oppressive sense. He said that he and his lab mates were terrorized by his mentor's daily stroll through the lab saying, "Where are the data?" But, even with the memory of that anxiety, he wants to make an effort to get around to the lab and see what people are up to in a nonthreatening way and not just stay sequestered in his office. I really enjoyed pondering Jeremy's success at running a lab without graduate students, an issue very close to my heart. I think that Jeremy's reflection on how to run a lab is best summed up by his funny, but also totally true sentiment that you, "Do the best you can and try not be a jerk."

CHAPTER 12

How to Have a Great Life and a Great Career

The Teeter-Totter of Work-Life Balance

by Ashleigh

Although I use the term *balance* in this chapter, I should start off by saying I am using it for lack of a better term. I know that there is no such thing as work-life balance. In fact, I argue that to be happy in both your personal and professional life, you should never be trying to balance both. You should be all in when you're working and all in when you're with family—being in the middle (i.e., trying to balance both at the same time) is actually what seems to cause the most problems. In this chapter I describe how Geoff and I have sought a happy medium between work and life outside of work, but I acknowledge that different people have different perspectives and hope that what I say here helps you either because you can incorporate some specific examples into your own life to help you feel like you aren't alone, or as a basis for conversations with your peers and mentors.

THE PENDULUM OF BEING PRESENT

I have felt as though the way to maintain balance is actually to manage your work and private life as a pendulum violently swinging from one extreme to the other. Specifically, if you are 100% present and engaged in where you are—completely engrossed in work when you are on campus or consumed by your family when you are at home—you end up being sufficiently productive at work and engaged with your family that you experience a satisfying balance between the two. However, if you are constantly checking your email and working on your lectures or manuscripts in between cooking dinner and playing with your kids, your work will suffer, and your family will feel short-changed. If on the other hand, you close your office door most of the time on campus so that you can really hammer out whatever needs done while you are there, you can head home with a good amount of work accomplished and a clear mind, ready for a break from work to focus on your family.

For most of my career, I have been in a different town than my husband with three kids. This has led to most of the responsibilities of raising kids to fall on my shoulders while maintaining my career. My students have joked about my kicking them out of my office when we are chatting, but that is exactly how I have maintained a career as well as a fulfilling family life. Although I enjoy my students, I limit my time socializing with them or chatting with colleagues so that I can focus on work while on campus and then leave on time to pick up my kids from school and be present with them for the rest of the night. I need to be all in at work so that I can leave when the kids get out of school and be all in at home.

YOUR FAMILY CAN BE YOUR '79 VOLKSWAGEN BUG OR YOUR CACTUS

The advice here is for everyone. Between the two of us, Geoff and I have a blended family with four children, so much of what I talk about is trying to maximize opportunities to get back to the kids. But don't feel forgotten about if you don't have kids. Your family can be that '79 Volkswagen you are restoring, the greyhounds you foster, or your favorite cactus. I don't judge, and I really support (and are mildly jealous of) your eccentric personal life.

MAXIMIZING YOUR DAILY RHYTHM

My whole family knows that I'm just no good after 9 p.m. I'm loving and energetic and will give them the shirt off my back and the shoes off my feet until 9 pm. I get grumpy and irritated and lose patience with my lovely cherubs after 9 p.m. Part of the reason I become impatient with my family at this time is that I start to feel like I'm not ready for the next day; maybe I don't have my lecture ready or I haven't gone over the article for lab meeting. That means that after 9 p.m. is a great time for me to retreat away from my family (who I have focused on since they got out of school and are now in bed) into reading the reviews for a manuscript I need to act on or to hide behind my laptop and prepare tomorrow's lecture. After 9 p.m. is also usually when emails stop coming, the world becomes quiet, and I can just be alone with my work.

My family is really good about this. My kids respond really well when I remind them what time it is and that I'm tired and no good anymore. They are kind and understanding and give me space. They've been trained to do this, though. I've spent years teetering on exclaiming the *I'm gonna lose it* mantra when 9:30 rolled around and they would request another cup of water or get out of bed to complain about their toenail hurting. You can do this too; be clear with your family about what you need and do it kindly and firmly, over and over again. And recognize that this is part of your daily rhythm—you are not a bad parent or an impatient person. You have reached your limit of undivided time you can give your family and you need to now shift gears (i.e., allow the pendulum to swing back to work).

From when the kids get out of school until 9 p.m., it's family time. I'm not perfect (I know, you're shocked) and I definitely still check my email on my phone sometimes or bring my laptop to multitask if I'm going to be sitting at swim lessons for an hour or basketball for 75 minutes. But otherwise, I know my 9 p.m. break is coming and I worked while they were at school and so I give them (and me) time for our family. I think that having my mostly undivided attention after school also helps my family respect my 9 p.m. cutoff. They know I've given them my all; I've played on the trampoline with them, made dinner, made slime with the neighbor kids, checked their homework, discussed the latest cruel list the kids are making at school (e.g., who drew the worse butterfly at recess—why are kids such jerks with their list making?!), read to them while they took a bath, etcetera. We have some great family time from 3:40 when I pick them up from school until 9 p.m. I need that time with them and I miss them when they are in an activity that hogs those hours that we get together. But by giving my all during those hours, I can focus on work guilt free the rest of the day (and night if I want to work after they go to bed).

I also have to admit that once my kids are older and less interested in being with me all the time, I envision stepping it up career-wise. I imagine that because I love research and writing, and there is always more I can be doing, I plan to be more productive when my kids are older, when I don't feel pulled to both be working and at home. This is one of my favorite things about my career. I've been allowed such a wonderful degree of flexibility that I don't regret how I spent any of the years before my kids hit elementary school age. I know that I spent as much time with them as possible while still maintaining a decent career and that as they stay after school for sports and clubs or grow to find summer camp equally or more appealing than hanging out with me, I will be able to become more productive at work. This is perhaps my absolute favorite thing about my career.

BE A GROWNUP ABOUT YOUR LISTS

One of the many benefits of being an academic is the flexible schedule. But if you have some drama in your personal life (as we all do), that can mean room for rumination. So unlike the mean children at my kid's schools (seriously, who taught them lists are meant for cruelty, like who has the worst haircut?!), I use a list like a grownup. I have a pad of paper on my desk where I jot down all the things I need to do, whenever they pop into my head. They might be small and unrelated to work (e.g., buy dog food) or they might be big (e.g. make progress on grant that is due in a month) but it allows me to take any moment that is otherwise unfilled and unaccounted for and do something productive rather than spend too much time on unproductive tasks. Geoff also feels really accomplished crossing items off a list. Later I will talk about how to survive this career when you so infrequently receive positive feedback; one of Geoff's tricks is to make a list and cross off the items. It fills him with a sense of accomplishment and positive feedback in a cycle he creates for himself.

KNOW YOUR BEST THINKING TIME

Geoff thinks of most his best experiment ideas while he's mowing the lawn. We used to have a very large garden in our backyard in East Nashville and he'd go out back to the garden and water for much longer than it should actually take. I'd tease him that he was trying to get away from our family. That was in part true, because he wanted some space, some sunshine, some birds chirping, as he thought about his grant, his manuscript, or a solution to a mentoring issue he was facing. I do some of my best thinking when I'm in the shower. I often give myself a little pep talk before I get into the shower about what I'm going to think about while I'm in there so I don't get off topic and spend that time ruminating about things beyond my control. Instead I start to plan out my next experiment, manuscript, or grant. Figure out your best thinking time, as unconventional as it may seem, and try to protect it.

THERAPY AS A TIME MANAGEMENT TOOL

Therapy is a useful time management tool for me. When I know that I have an upcoming appointment to process something that is going on in my personal life, I can essentially table it. When I know that I'm going to confront and deal with it later, I can focus on work. To this end, I often open my calendar to the date and time of my upcoming appointment and jot down a quick reminder about what I want to discuss with my therapist. This offloading of the troubling issue to my calendar means that I no longer need to remember it, giving me some peace of mind to focus on my work.

WHEN HAVE YOU PUT IN ENOUGH TIME?

I hear my academic friends belabor this issue all the time. One of the potentially serious challenges of being an academic is that there is literally always more you could be doing at work. You could be writing another grant, you could be reviewing more articles, you could be reading more articles, you could be designing more experiments, you could be programing more experiments, you could be analyzing the data that was collected that day, you could put in more hours mentoring your students, the list goes on forever. So how can you pick up your kids from school and bounce on the trampoline with them guilt free? You decide to, that's how. You decide when enough is enough. After years of trying to strike a teaching and research balance that felt good to me, I decided that, while my kids are young, I'm basically a two-paper-a-year kind of professor. I have a full teaching load (3–4 classes per semester) and run a lab with 10–15 undergraduate students each semester. My appointments are much more about teaching than research, but since I love the research, I've had to figure out how to strike a balance between what I want to do and what I'm paid to do. I'm mostly paid to teach, so once I have my classes covered (e.g., I've put in my prep time on my lecture, written the next exam, responded to student emails in a reasonable amount of time), then I spend any extra time I have on research. It could be an entire day that I only spend on research or I might go an entire week with absolutely no time for research. That effort and balance has resulted in averaging the publication of two papers a year in peer-reviewed journals. I'm happy with that outcome, it works well for me, and it's what I can get done in the time I'm willing to and interested in allotting to my work life. Could I get more done? Of course I could. Am I willing to put in more time and take it away from my family? No way, not at the age my kids are and with their desire to be with me. Someday when they aren't home right after school I'll put in more hours at work but for now, I'm just not willing to feel bad about what I can get done in order to be there to pick up my kids when the dismissal bell rings at school.

The reverse lifestyle might work better for you. I have a friend who needs to maximize after-school care for her kids in order to feel like she can be a good mom at night. She's pulled into the parking lot at daycare and read a manuscript in her car because she still had time before they closed. Feminism is about the right to choose the life that works for you but unfortunately mothers in particular feel lots of pressure to maximize time with their kids, often that pressure comes at the detriment of their kids. Don't let anyone guilt you into thinking that there is one way to have a career and a family. You need to do what works best for you because if you're taking care of yourself, your career, and your family will reap the benefit.

After you decide on your boundaries you'll know when you've put in enough time at work. If you are cracking under the pressure of writing grants and publishing at your R1 job, and it leaves you miserable and unable to see your family, it might be time for a career change. If you have healthy boundaries and have a job you can manage in the time you're willing to allot to your job, then I think you've hit the jackpot.

If you regularly bemoan that it's impossible to know when you've put in enough hours, step back and ask yourself (or your department chair, or a trusted faculty mentor) whether you are right on track for tenure (or whatever your goal might be). If you are way ahead of schedule, then take a deep breath and scale back—you are fine! If you are right on track, or behind, but feel like you're already giving more than you have, then I'd say you might be in the wrong position. Ask for that feedback from people who know you and the requirements of your specific position at your specific institution well. Go to your chair or senior members of your department and ask for honest feedback on the likelihood that you will earn tenure and be ready to receive what's said with an open mind.

DON'T LET YOUR EMAIL OWN YOU

Setting up a method to deal with email is central to a chapter on the teeter-totter of the work-life balance. There is so much advice out there about how to better handle email, so here I just mention what I know works for us and encourage you to find a rhythm of your own.

If a large portion of your email is from student communication because the majority of your appointment is teaching related, setting some expectations about email communication on your syllabus can be important. I came to this realization during my first faculty job when I heard students complain about another professor not answering email within an hour or two. They would say they know she gets email on her phone and so they knew she had received their email and were irritated that they hadn't heard back. This was back in the day before everyone got email on their phones, so that declaration sounds stranger now, but it struck me as so violating and inconsistent with the expectations that should be placed on me and my colleagues. As mentioned in Chapter 8, now I have a section on my syllabus that clearly states when students can expect to hear back from me and what to do if they don't hear back in that amount of time.

My boundaries are that students can expect that I saw their email within 24 hours if it's during the week and 48 hours if it's the weekend. I tell them that sometimes I check my email on my phone and want to type up a thorough response, so I put away my phone and plan to respond later. But by the time I get to my computer, I have forgotten about their email, and will neglect to respond. So if 24 (or 48 hours) have passed, the email may have fallen far enough down in my inbox that they won't hear back from me without resending it with something in the subject line that will catch my attention or clarify that it's a second attempt. It really helps me get serious about responding if a student has had to make a second attempt and it also lowers my stress over not going back through emails to make sure I've responded. I'm not one of those people who will respond to an email that was sent a few days ago. I never comb back through my inbox to make sure I'm not missing anything. I think that's a waste of time and assume that if it's important and I happened to miss it, they'll email me again.

Another method to managing email is to literally only open your email once a day. That way you train people that they will not get a fast response from you and they should reserve emailing you for situations in which it's really required. If you don't respond to emails right away, often the issue magically resolves itself and the sender finds another avenue to get their answer. This encourages trainees to independently problem solve.

Some other tips on handling emails include the following: Don't engage in mass emails, exit your email application when you are writing, and do not have emails pushed to your phone automatically, especially when you are with family and especially on weekends. Recently I've used Slack for all my lab communication, which allows me to compartmentalize and prioritize my research.

MEETINGS

If someone wants to meet with you, ask for an agenda and give them some tasks to do before the meeting. Tell a graduate student that they need to analyze some data and send you a figure before you'll meet with them regarding an ongoing project. Tell the undergraduate who has questions about material on the upcoming exam to send you a completed study guide, explaining that to maximize your meeting time, they need to have put in the studying hours first. This allows you to spend less time on pointless tasks (i.e., aimless meetings) and more time maximizing your work so you can go home and enjoy the rest of your life.

DEALING WITH INFREQUENT FEEDBACK

One of the best things about life as an academic can also be one of the worst; you're basically your own boss. When you're basically your own boss, you don't have anyone breathing down your neck, demanding you come in at a certain time and leave at a certain time, providing constant feedback through micromanagement, et cetera. However, one of the hardest parts of being your own boss is that you don't get much feedback at all and when you do, it's probably negative. Some of the only feedback we get if we have a research-focused career are reviews on our submitted manuscripts or if we have a teaching-focused career, student evaluations of our teaching. Both of these kinds of reviews can be harsh. One of the best cures for negative teaching evaluations is sharing them with colleagues so you don't feel alone.

Laughing about the cruel, misogynistic, and irrelevant things that students write on teaching evaluations can help, but there are some potentially less cynical ways of dealing with infrequent and negative feedback. One is to buy a stationary box in which you plan to keep all the kind comments you get from colleagues and students. Over the years, collect the thank-you notes you get from a variety of sources and put them in this box. When you get a congratulatory email or a positive teaching evaluation, print it out and put it in the box. I have a friend who prints out positive emails and evaluations and hangs them on a bulletin board above her desk. Recognize that academic work can be a long, lonely road (which most of us enjoy), and you need to settle into a rhythm that works for you. I keep that box of thank-you notes next to my home desk, and I have more notes on a bulletin board in my campus office, both in full view every time I sit down to work. It's the motivation I need—find yours.

WHEN IS A GOOD TIME TO HAVE KIDS?

There are two answers to this question that I whole-heartedly endorse. First, there is no good time to have kids. Second, every time is a good time to have kids. Both of these are true sentiments. I had two of my kids in grad school and one during my first tenure-track position. Each experience was easier for some reasons and harder for others. Have kids when you want to, and everything will fall into place; or it won't and you'll know you need to make some life changes.

When you have kids in grad school, you are able to take advantage of the flexibility that a student has. There are not many meetings you are required to attend, there aren't many classes you need to take (especially if you have the kid when you're ABD), you probably have an undergraduate to help you collect data so you don't have to do that on your own anymore, et cetera. The struggle is usually money. It's hard to afford kids on a grad school budget, mostly because daycare is so expensive. The upside is that you have fewer events to attend (see previous points) and you have lots of friends (other grad students) who likely don't have kids and are more than happy to help out (so they can decide if they want to have kids of their own and be covered in boogers and vomit someday too). My friends in grad school helped me out with babysitting so I could attend classes and meetings with my advisor. I spent a lot of time working from home or nursing in my office on campus. Another bonus, the institution where I went to graduate school offered me PAID maternity leave from my position as a TA. It turned out that my maternity leave support in graduate school were better than during my first faculty job.

When you have kids during a faculty appointment, the huge advantage is that you are making significantly more money to pay for the expenses that go along with having a baby. I will add that you can get by with very little for your child (assuming you have health insurance); not being able to buy the top of the line crib mattress because you're a poor graduate student isn't really a curse and is in some ways a blessing (because you don't need that crap anyway). Many folks find being a faculty member less of an ill-defined problem than graduate school. In graduate school, many people feel like they are constantly guessing if their advisor and program is satisfied with the amount of work getting done and they have the looming concern of landing their first job, which might make juggling having a child during that stage of the career harder than as a faculty member. As a faculty member, you might feel more self-confident in setting an away message on your email, putting up boundaries about not checking email during family time, and extending your tenure clock to allow time for the baby.

Overall, there are pros and cons to having kids regardless of where you are in your career, so I say just do it when the time is right in your personal life and let the chips fall where they may.

MATERNITY LEAVE AND THE TENURE CLOCK

As hard as it is to believe, some institutions still do not offer paid maternity leave. When I had my youngest child, my faculty appointment was at an institution that did not offer paid maternity leave. If you are living separately from your spouse and therefore each supporting a household, are a single parent, or are the primary breadwinner in your household, taking the unpaid FMLA (family medical leave act) for maternity leave may not be an option. I solved this issue by unapologetically bringing my baby with me to work. The first few months of his life, my baby was cuddled up against my chest in a baby carrier while I lectured or lying on a blanket in my office when I wasn't nursing him. It was hot and sweaty and difficult, but I also didn't have to worry about him being old enough for daycare or deal with pumping breast milk. One time I lectured with his poop on my hand because he had a blowout in the baby carrier during class and I tried to not let the students know about it. The struggle is real, but it can be done.

One common question people face is whether it makes sense to add time to their tenure clock. The stage during which many folks have children is pre-tenure, which raises the issue of adding time to the tenure clock. But you might add to the tenure clock for other reasons, for example, taking a semester to care for an ill family member. It is often debated whether this is a good idea or not. The school of thought behind not adding time to the tenure clock is that people will increase their expectations because you took more time or in some other way look down upon you for taking the extra time. However, Geoff and I disagree with this sentiment because people want you to succeed. The folks in your department offered you the position because they want you there. Keep in mind that they have invested a tenure-track line in you, potentially $100,000 USD in your salary and benefits each year, and possibly $500,000 USD in startup funds. So by the time you go up for tenure, they have invested a million dollars in you. They lose that money if you don't get tenure. They lose more than that money too, because their ability to recruit faculty will be hurt if you don't get tenure and there is any suspicion that the reason was because you added time to your tenure clock due to major life events. Institutions want to be able to tell potential faculty members that they are family friendly. They want to tout that you can have a good work-life balance and still get tenure at their institution. Thus, they are unlikely to give you the short end of the stick.

Generally speaking, Geoff and I both suggest adding the time to your clock when in question. The reality is that it only helps enhance your tenure package. It allows for more time for those submitted publications to advance through the peer-review process, to continue data collection in your lab if you have students working when you aren't there, gets you and your students additional time for conference submissions and presentations, et cetera. The idea that there might be some harm in terms of bias against you does not appear to outweigh the good.

SINGLE PARENTS

We consider ourselves single parents in the sense that we have lived apart for most of our relationship and juggled kids without the day-to-day help of an adult partner. We are huge advocates for improving the quality of life of single parents by bringing attention to their plight. First, oh wonderful single parents and allies, say no to meetings at inappropriate times (e.g., 5 p.m.), even when you can make it, just to make a point about how inappropriate the time is. This is especially vital when you are established enough or comfortable enough to do so. Advocate for those 4 p.m. brown bag talks to be moved to noon or some time that single parents don't need to be picking their kids up from school or daycare or driving them to swim team practice. Being on campus all day should be enough time for scheduled talks.

Let's just drop a line about bringing your kid to class. Many of us have stories of that one student who provided us with feedback that they didn't realize how anyone could juggle a career and a family until that day when our kid had a temperature and couldn't go to school or daycare and we had to bring him to class. Will some students say it was distracting (even if the kid is asleep on your chest)? Yes, someone will probably complain. But will you model how to be a parent and have a career, which will inspire someone too? Yes, you will. Be willing to take some heat from one negative student to inspire another student. We've all done it! Keep some snacks, coloring books, markers, diapers, and toys in your office and just do what you need to do. I have colleagues who have their kids dropped off at the bus stop on campus. I have a colleague whose child navigates the city bus system after school and has her own key to her dad's office to let herself in when she arrives on campus. I have colleagues who send their kids to the student union to get ice cream after school. My daughter used to ride her bike in a loop between my office, an empty storage area, the hallway, and my colleague's office during my office hours. My kids know which faculty members have candy bowls in their offices and can wander the building unsupervised checking on the candy bowls while I work. If you have kids, bring them to work, it'll be ok. If you don't have kids, support your colleagues who do have kids.

SALARY

In previous chapters we discussed negotiating for your salary. Here I discuss the multiple ways to increase your income because more money can make your life outside of work much more enjoyable. I also describe some of the ways we have increased our income and the ways it happened for us in the hope that it helps you create opportunities for yourself.

SERVING ON A PEER-REVIEWED JOURNAL'S EDITORIAL BOARD

Working your way up an editorial board to a paid position, such as an associate editor, can result in an extra $4,000-$12,000 USD per year, depending on the journal, and more if you are the head editor. That's just a rough estimate based on our experience, but you get the idea. There are a number of ways you can end up on these boards. The first is to serve as a peer reviewer. I don't want you to be overwhelmed with reviews, especially if you are pre-tenure, but you should say yes to the reviews that you can do efficiently and effectively. Writing good reviews will sometimes get you invited to be a consulting editor at a journal. That is not a paid position, but it goes on your CV. Then once an associate editor position opens up, you are more likely to be invited to serve. I served as a consulting editor for a journal and when I had a chance to speak with the editor, I explicitly expressed interested in becoming an associate editor. I think that put me on his radar as someone interest in the position when the time came that there was an opening. It is also not uncommon for the editor to ask the current associate editors who they might recommend when someone steps down and they have a position to fill. You should conduct yourself in a manner that puts you on their radar. For example, be intelligent but not belligerent at conferences, write thoughtful and timely reviews, and be known as a pleasant person who gives objective, fair, constructive, and measured feedback. Finally, if you are interested in serving on an editorial board, tell people. Let your colleagues know so that in the event that there is an opening, you will be on their mind as someone to recommend.

WRITE A BOOK OR TEXTBOOK

Use the unique approaches you have developed for teaching courses over time, or your unique pedagogical outlook, or your unique expertise, and propose a book that isn't already on the market. There are lots of ways in which the current textbook model, and therefore the current publishing model, does not work for today's undergraduate students. There is a lot of room for your creativity to collide with the needs in the market and book publishers are generally very receptive to listening.

If you work at a larger university and your class enrollment is very high, it makes a big difference to some publishers because those are essentially guaranteed sales. Inspiration for my first edited anthology came from my experience of incorporating journal articles into my classroom over the years. However, the reality of why it ended up happening is that an acquisitions editor from a publisher came to Ohio State and asked for meetings with people teaching high-enrollment classes. At the time I met with the folks from the publisher, I was a lecturer, and we embarked on a great relationship and a wonderful (if I do say so myself) finished product. You do not have to be famous (or even have a tenure-track title) or have a wildly successful research career to get a book deal. You just need to be thoughtful about teaching.

The book you are holding in your hands is the result of Geoff and I deciding that we should just write down all the things we have discussed with each other over the years. We wrote a book proposal and submitted it to about four or five publishers. Two sent it out for review, one was particularly responsive, and the rest is history. We used a colleague's book proposal template. Although there are many templates you can find online, if you are thinking of writing a book proposal, it would be wise to get a successful proposal from a friend or colleague as a model.

TEACHING EXTRA CLASSES

Different institutions handle overload courses differently, but there might be an opportunity for you to pick up an extra class, beyond your current teaching load, for extra pay. Another opportunity I have embraced almost every summer is to teach summer school courses. This is actually a very fun way to travel to a new place, experience a different institution, meet different types of students, or enhance your CV. I have a friend who teaches Maymester at her home institution for extra money and then a summer session course at a different institution because her home institution caps how much she can make teaching overloads at a percentage of her pay. That means she can't make much additional money if she teaches summer school at her home institution, so she teaches summer session somewhere else across town. Inquire about any such rules at your school, and if they cap your income, check out the other schools in your area. Institutions usually don't post these positions, so just email the chair in August or September expressing your interest in the opportunity to offer a summer school course in their department, attach your CV, list some courses you are qualified to teach in the body of the email, and offer some references or a teaching evaluation.

SUMMER SALARY OR GRANT FUNDS

Enhancing your salary with summer income from grants is available to many of us, even without a wildly successful research career. Of course, if you have a large grant, you can draw summer salary to augment your nine-month salary. However, for the rest of us, there are many small grants that provide some kind of income for working with undergraduate students during the summer. For example, the APA summer research grants come with a stipend for faculty members to use to enhance their research, which could mean a new laptop that you now don't have to buy with your own money, or subject compensation so you can get pilot data collected for a grant, which will in turn fund your summer salary. Many Psi Chi grants come with funds for faculty members who oversee the research. Check into the possible funding sources available to your students. Often they come with bonuses for faculty members as thanks for working with undergraduate students.

STUDY SECTION

The government determines who is going to get their research funded by asking experts in the field to peer-review grants. Usually being a peer reviewer requires reading, reviewing, scoring, and submitting comments online for a subset of the study section's grants. Then members of study section fly to a hotel where they meet for 1–2 days to discuss the grants and potentially debate the scores. In addition to having expenses paid for these trips, the folks who serve on study section are usually paid a few hundred dollars for this task. The way to get invited to study section is usually to land some of these big grants, so there isn't much you can do to get on study section other than do great work and write good grants. However, some other grant reviewing opportunities do accept applications such as the National Science Foundation. You can fill out an application to be a reviewer for the Graduate Research Fellowship Program (GRFP) online, which in 2018 paid $200 per day for each virtual one-hour panel session attended[1].

SPEAKER HONORARIUM

When a speaker is invited to talk at an institution or conference, they are typically awarded an honorarium on top of having their expenses paid. These are usually small, like a few hundred dollars, but they are still a nice little bonus. The honorarium is usually larger if the speaker has to make an international trip to present. In our experience, the prestige of the institution usually has very little to do with the size of the honorarium. One nice thing to do is to invite your friends to present at your institution with whatever funds you can secure from your department. It is obviously a great opportunity for your students to have them come speak but it also does your friend the favor of padding their CV and getting them a little extra money. Hopefully, they will get the hint and return the favor by extending you an invitation to their institution.

SERVICE POSITIONS

Many service positions in your department and across the university come with small bonuses. Serving as director of graduate studies or department chair sometimes comes with a teaching release or additional research funds (which can in turn generate summer salary if you spend the money collecting pilot data for a grant). Other small service opportunities might also offer some unexpected financial reimbursement. For example, I worked at an institution where they asked faculty to spend a few days doing summer advising for the incoming first-year students and they paid us a little bonus as a thank you at the end of the summer that I did not know was coming. The main way to get these service positions is to express interest in them to your chair. Recognize that they can take up a lot of time and make sure you want to do them. If you get an offer to serve in a position, try to negotiate for more than whatever you are being offered.

JUST SAY NO

The best of intentioned people will walk all over you if you let them. To me, the work-life balance question really boils down to knowing you've done enough at work so you can enjoy your life outside of work. That means saying no when you've hit your limit of doing enough. This is hard to figure out when you are new in any position, at any stage in your career. Look to your senior colleagues or a trusted faculty mentor to tell

---

1 That application can be found here: https://nsfgrfp.org/panelists.

you. Let them know that in order to be effective at work, you need to also have some down time, and you are trying to figure out what requests you can reject. Make a list of the requests you receive over a few weeks and take it to this person. Ask them which of those requests you should always accept and which you can reject. Here are some big ones for me:

- Requests to review manuscripts. I believe that you should review three papers for every one paper that you submit for publication. You need to serve the field if you expect other people to review your work when you submit it. But the quality of your work (and life) will go down if you say yes to everything. That's the 3-for-1 reviewing rule. This is the case until you're on an editorial board and then you can say no to all other requests that aren't for that journal. I don't say no to everything else when I am on an editorial board, but I am very selective and turn down 90% of requests, citing that I need my time to be available for the journal for which I work. That way I can act quickly on papers once the reviews are in and serve the field that way.
- A request for a meeting is a request. You can usually say no. It's worth repeating here that if someone asks you to meet and their rank is lower than yours, you should give them some assignments first. This could be true for other faculty in addition to students. For example, tell them you will meet about their new project proposal after they send you a figure describing the paradigm and an abstract they intend to revise for a conference submission. Or you will meet to discuss their data after they've run multiple analyses you request and emailed you the figure. If their rank is at or above yours, it is still ok to ask for an agenda. If the agenda seems like something that can be accomplished in an email rather than a meeting, just say so. Be clear about when you are and are not available. Never agree to meetings that are outside normal business hours—even if you are free, you set a bad expectation for folks who have family obligations at those times and can't meet then.

WORKING FROM HOME

I suggest figuring out where you stand on working from home. Some people have a lot of success working from home and other people find it insufferable. For those who have young kids, some find that trying to explain to them that you are working and need to be left alone is just too strange of a concept for a child to understand. I suggest getting a great pair of noise-cancelling headphones and settling into a favorite café while a babysitter is at home with the kids. Even if you don't have young kids at home, some people find that the laundry and the dirty dishes and the endless domestic tasks or distractions at home interfere too greatly with their work. On the other hand, some people find that being on campus is too much of an open invitation for intrusion and interruptions from colleagues and students. One institution I worked at requested that we post weekly schedules on our doors and I blocked out writing hours where I claimed to be off-campus writing. I might have actually been hiding in my office, but I made it clear that I was unavailable and nobody knocked on my door. The reason I raise this issue here is that finding a way to maximize your work hours allows you to maximize your family hours, too.

HOW TO DO A GEOGRAPHICALLY TARGETED JOB SEARCH

One of the most common situations we hear about from folks in the market for a job is that they want to move to a specific town. This, of course, is tough. Not only does it require that one of the institutions in the town is hiring, but also that they pick you out of all the available candidates who have applied for the position. What should you do if you want to move to a specific town? You should not move without a job offer, work remotely at your previous institution, and hope to get your foot in the door. Why not? Well, because while you have moved and are working remotely, you have left the intellectual community that stimulates your work. This results in reduced productivity and then other potential applicants can whiz past you to create a longer, more impressive list of publications, service experience, pilot data for their grant, teaching opportunities, et cetera.

So what should you do? You should definitely email the chair of the department where you want to move and tell them who you are and what you want. Tell them you are from the area and want to return to be near family now that you have kids. Don't say that if it's not true, but you get the idea. Tell them you are very interested in ending up in their geographic area. Then follow up in about six months if you haven't heard anything, maybe telling them you'd love to give a talk next time you're in town. Just don't ask them to meet with you—that's needy and annoying. Approach other people in the institution's department at conferences and ask them how they like working there and express your interest in ultimately ending up at that same location. There are lots of things you can do to express your interest while you continue to grow your CV.

CONTRIBUTING TO A FAMILY-FRIENDLY DEPARTMENT

Whether you have kids or not, whether you are 35 or 65, you can help contribute to your department's work-life balance. Here are some ideas that I wish people had done for me, and that people did do for me and for which I am eternally grateful. Requests for attention to work-life issues have more power if they are not coming from the faculty member who needs the immediate help, so please speak up for the rest of us. I have mentioned some of these issues before, but they are worth reiterating because I want change, people!

- Do not have speakers talk at the end of the day. A speaker having breakfast with some faculty and then giving a talk at 10 a.m. rather than at 4 p.m., when everyone is sleepy and grumpy, makes the visits better for the speaker and better for the folks with family.
- When you can, take less appealing time slots to meet with the speakers. If your family is out of town, if your kids are at summer camp, if you have a shared parenting situation and it's not your night, if your kids are in college, et cetera, help your colleagues with kids or otherwise complicated lives by letting them have first dibs on meeting times.
- If your institution does not have paid maternity leave, fight for it, especially if you are tenured and don't need the leave yourself. Not having paid maternity leave is a disgusting affront to families, especially women.
- Do everything you can to welcome kids on campus. Tell faculty members who bring their kids that you are glad to see them on campus. When I worked in Indiana, a PR representative stopped me in the hallway when I was carrying my daughter Paige's bike up the stairs with my infant son, Henry, in my other arm. She said she'd seen me do this many times and wanted to take a picture of me juggling the kids. It appeared in the school magazine with a lovely message about the institution being family friendly. I look back on that photograph and those days so fondly. That woman didn't have to approach me that day, or ask to take our photo, or be so supportive of my juggling act as a single parent. The fact that she did meant the world to me and I think about it often.
- Be honest with your away messages. If you are on vacation with your family be open and honest about it. If you are on maternity leave say so in your away message. If you are traveling with a band in which you play the ukulele sing its praises in your away message! This makes it less taboo for other junior faculty to feel like they can be open about having lives too. It also sets a great example for your students that they can have both a career and a life someday. When I get an away message from someone asserting they are on family leave, I respond congratulating them on taking the time for themselves. We need to be as supportive as possible of one another.
- Bring awareness to your institution about research showing that gender neutral family leave policies actually benefit fathers and hurt mothers. The idea is essentially that a women's physical experience of carrying the child, giving birth to the child, nursing the child, and being less likely to pass off childrearing duties means that she actually spends this leave on nonwork tasks. However, because men's physical condition is unchanged following childbirth, men can spend some of the

time-off working, using family leave to enhance their tenure package. The result is that men who take family leave and add time to their tenure clock are more likely to increase productivity and when they get tenure they essentially raise the bar for expectations placed on everyone else, including women.
- Provide support to parents by allowing them to select the family-friendly course times. Teaching a class at 8 a.m. or 4 p.m. usually does not work for parents who have kids to get off to school in the morning or shuffle to soccer practice, horseback riding, or violin lessons after school. If your registrar is putting pressure on your department to offer an evening class, please do not ask someone with young children to teach that class.

Ashleigh's Interview with Caitlin (Hilliard) Hilverman, PhD

caitlin-hilverman.com

Caitlin (Hilliard) Hilverman, PhD, is a postdoctoral research fellow in the Department of Hearing and Speech Sciences at Vanderbilt University Medical Center. She received her PhD from the Department of Psychology at the University of Iowa in 2016. Caitie is a mother to Poppy and wife to Ron Hilverman. She has been well funded throughout graduate school and into her postdoc, has numerous publications, strong collaborators, and is well respected in the field. I interviewed Caitie for this chapter because she is thoughtful, wise, and honest. Caitie has recently juggled numerous major life changes. She is new to marriage and motherhood, and moved to Nashville for her postdoctoral position at Vanderbilt immediately after graduate school, while pregnant. I knew Caitie would have some realistic, genuine feedback on my suggestions above and be willing to provide some very specific examples of her attempts to maintain sanity, productivity, and a happy life outside the lab.

BOUNDARIES

Caitie had so much to teach me about setting boundaries. She suggested making it a habit to begin each semester reestablishing boundaries based on what you need at that time, incorporating the issues you encountered and strategies you learned from the previous semester. For example, after a particularly burdensome and overwhelming teaching assignment, Caitie instituted policies with her class that she would not answer emails on the weekends, would not help with stats over the phone, and would not answer any course-related questions in the hallway. Caitie initially felt like it would come across negatively to assert these boundaries, but a number of faculty members commented that it was healthy, mature, and impressive.

After some time in graduate school, Caitie was no longer willing to help with statistics in return for merely being thanked in the acknowledgements. She learned that as she became better at time management and more proficient in her own flourishing research career, ten hours of her time analyzing someone else's data just was not worth being thanked in the acknowledgements section of their paper. This was learned the hard way, after she recognized, *I'm not enjoying this anymore. People senior to me (e.g., faculty members) are asking me to do something for them, and I'm not getting anything from it*. I really admire Caitie's reflection on how important it was for her to stick up for herself but how it can also be difficult when you are doing so with people who have power over you, like the faculty in a graduate program. Caitie stopped giving away so much of her time for free. For example, when asked to help with analyses on someone's project, she responded by saying that she would be happy to help, but she needed to be an author on the paper to make it worth her time. Or if she was asked by a close friend to help with analyses for a project outside her area, she made arrangements for them to give her something in return so that she didn't feel like she was prioritizing others' work over her own.

Caitie also reiterated the importance of setting email response expectations. When she was on maternity leave, she told the students who were working on her projects that she'd email them every Friday. Caitie still checked her email throughout the week, but she would not respond unless it was urgent. She would get about eight emails from the students throughout the week and she would work on an email response slowly, but would not send them until Friday.

This concept of not setting the expectation that you are going to respond to emails right away can also be aided by apps like Boomerang, which lets you change the time of your response to an email to be, for example, Monday at 9 a.m. That way, even if you do work on the weekends, you are not training people to expect a weekend response. Caitie said when she started using Boomerang to delay sending emails she'd worked on over the weekend until Monday morning, she started to get less email over the weekend. She also uses Boomerang to resend emails to herself later when she will have time to respond to them, say on Thursday at 3 p.m.

Caitie had a meeting with her faculty mentor in graduate school to set explicit boundaries. She found this to be a very professional and useful approach to their relationship because they were becoming good friends, which felt like unfair territory for a faculty member and their graduate student to be in. The meeting outlined issues such as how often it was reasonable and fair to meet, how to deal with questions that came up outside the weekly meeting time, how to reach out if she had questions outside of their weekly meeting time, and also set some personal boundaries. Caitie left this meeting knowing that she had a weekly meeting time that she was required to come to with a list of things to discuss. Caitie could email her mentor but if she didn't hear back, then she knew her mentor thought it could wait until their meeting time. Similarly, Caitie was welcome to drop by the mentor's office but if Caitie popped her head in and her mentor shook her head no, this was a signal that her mentor was up against a grant deadline or something similar and Caitie should not interrupt her. They even set boundaries about Caitie not babysitting for her mentor's kids, despite the fact that Caitie had been a nanny for a long time and frequently babysat other people's kids in the department.

Finally, Caitie had some suggestions about setting boundaries at home. Her husband and her daughter's caregivers know that when Caitie is in her home office with the door shut, anyone who wants access to her has to knock and wait for a verbal response before they come in. Caitie also reserves the right not to answer a knock if she is engrossed in work and since everyone knows that she might not answer, nobody takes it personally. This is consistent with other suggestions I made above about explaining your needs to your family in advance and that pre-planning usually creates situations that go smoothly.

MORALE FOLDER

Caitie agreed with keeping a folder for the infrequent, positive feedback received in academia. She has a physical folder with copies of positive emails, accepted funding decisions, graduate school transcripts, annual graduate school letters that assessed her progress in the program, screen shots of NIH comments saying a grant was getting funded, an NSF letter saying she getting funding, a letter from a friend congratulating her on running a marathon, and other items that she can flip through when she gets a bad review and needs a little reminder of why she likes research. This file has grown to include personal feedback too, such as some wedding cards in which friends wrote some really nice things about her relationship with her husband, Ron. Caitie also has a folder on her computer that can be easily accessed from home or work. She called this file a self-esteem folder. It includes things like a screenshot of a personalized birthday rap someone wrote for her on Facebook one year and a flattering email from a friend. These little boosts are always at her fingertips and can help her actively turn her day around if she gets a negative review or bad news about a grant she's worked long and hard on.

THE GOOD LIFE CHECKLIST

Caitie periodically makes a good life checklist. This list serves as a gentle reminder to do some things daily that she knows contribute to her feeling whole and fulfilled. One of the lists on her computer included love on Ron (her husband), floss, walk the dog, read one article, make the bed, and get 8+ hours of sleep. She knows that the more items she checks off in a day, the better she will feel. This is my favorite thing Caitie taught me and here is why: flossing always feels like a luxury to me. I've never flossed as much as I should have and you know what my excuse has always been? I told myself I didn't have time! *Who do I think I am?! How long do I think flossing takes?!* Having that extra nudge of a good life checklist and putting such simple yet important things on the list strikes me as a wonderful, reassuring bit of advice.

SELF-CARE

Caitie suggested we talk about self-care in this chapter but not in the way you expect to hear about self-care. She commented that so much of what we see about self-care is about bubble baths and #wineontuesday. But self-care is not just getting massages; it can be something as nurturing as deciding not to set an alarm and stay in bed on a day when you feel like you're getting sick or scheduling a visit with your primary care provider. This was another great piece of advice for me because Caitie so clearly put into words how I feel about getting my mammogram. I feel like I'm taking such good care of myself when I get my mammogram that I take a selfie in the waiting room and send it to some of my closest girlfriends asking them if their next mammogram is scheduled. I feel like I'm loving on myself and my friends when I snap that picture of me in my robe waiting for my appointment and reminding people I love to make their own appointments.

A BETTER WAY TO ORGANIZE YOUR TO-DO LIST

Caitie showed me that she organizes her to-do lists on her computer by category. Here's the cool thing, there are three categories: self-care (her current list included scheduling a visit with her primary care physician, scheduling dentist appointment, picking out a yoga class, waxing her eyebrows, scheduling a 10K, buying eyeliner), work stuff (Caitie keeps this list detailed at the level of "write two paragraphs of the discussion section of a particular manuscript," or "return a colleague's email")," and fun projects (for Caitie these things mostly involved her daughter Poppy like "find a lobster onesie for Poppy" or "work on baby photo album"). She pointed out that having a fun projects to-do list allows her to maintain a list of things she would like to have (e.g., she said she knows she'd really like to have a physical photo album of Poppy's first year) but that she wouldn't prioritize if she didn't have it on a to-do list.

Caitie also said she agrees with Geoff about the satisfaction of crossing things off a list, even when it's on the computer; she crossed out the item about emailing back her colleague it in front of me and smiled satisfactorily because she had done it earlier. The most interesting part of Caitie's to-do lists was actually the level of detail she included on her work list. I have a grant I should be working on, but I keep getting stuck when I work on it and then making a reminder that I need to work on the grant. But I don't write down exactly what I'm stuck on or should be thinking about with regard to the grant, so I haven't made much progress over the last few weeks because "work on grant" is way too vague for a to-do list to be helpful. Caitie taught me that I need to be much more specific about the next actionable work items on my to-do lists.

Glossary

ABD – All But Dissertation. This is the term for the status in graduate school after you have completed all of your class requirements and passed your qualifying examinations. The term describes the fact that the only requirement for your PhD that remains is for you to complete and defend your dissertation.

Assistant, Associate, Full Professor – There are three ranks for tenure-track professors (see below for the definitions of tenure and tenure track). An assistant professor is the rank for people who are in the first phase of their career. Typically, people are promoted to associate professor after being granted tenure, however, some universities promote people to the associate level prior to tenure decisions. Whether this is the procedure at a given university is a good question to ask when interviewing. After the associate level, people are promoted to the rank of full professor and these are typically the most senior people in a department.

Glossy Magazines – This is a slang term to refer to the journals with particularly high citation rates, such as *Nature*, *Science*, *Cell*, and perhaps the *Proceedings of the National Academy of Sciences*, that publish research across scientific fields. The term comes from the fact that these journals typically have slick, glossy covers and are sold as magazines (i.e., *Science*, *Nature*, *Cell*). Publishing in these journals is considered prestigious, however, getting your papers into these journals can be extremely difficult.

HBCU – Historically Black Colleges and Universities. This is a label used to describe a group of schools that served black students in the United States prior to The Civil Rights Act in 1964. Prior to 1964 it was possible for institutions of higher learning to exclude students based on race and these colleges and universities were created to educate black Americans. Most of these schools are in the southern United States, although not all, as they also exist in states like Pennsylvania and Ohio. HBCUs still serve primarily black students but include students of all races.

Impact Factor – Represents the average number of citations earned by papers published in that journal.

K99/R00 – These are grants from the National Institutes of Health that are known as *Pathway to Independence Awards*. These grants provide support for two additional years of postdoctoral training during the K99 phase, including research funds and salary, and three years of funding in the R00 phase that begins once a tenure-track job is started. The K99 phase is the first two years of the 5-year grant, and the R00 phase is the last three years.

NIH – National Institutes of Health (https://www.nih.gov/). This is part of the Public Health Service (PHS), which itself is part of the Department of Health and Human Services (DHHS). The Public Health Service is the government agency in the United States that is charged the nation's health and safety. The National Institutes of Health is a collection of 21 institutes and centers (https://www.nih.gov/institutes-nih/list-nih-institutes-centers-offices) that focus on preventing disease and promoting health. For example, the National Cancer Institute's (NCI's) mission is to conduct research related to treating and curing different types of cancer. The NIH is the largest funder of basic and applied research in psychology and neuroscience. The NIH funds extramural research at institutions nation-wide and also has intramural researchers who carry its mission of NIH by conducting research on a campus in the Washington D.C. area (Bethesda, MD).

NRSA – National Research Service Award. These are training grants from the National Institutes of Health to support graduate training (known as F31 grants), and postdoctoral training (known as F32 grants). These grants pay your salary during your training, as well as including a small pool of funds for educational expenses. Officially, there are two types of NRSA grants. One is the institutional NRSA variety that is granted to a university to fund a group of students and postdocs studying a common topic (these are known as T32 grants). The other variety is the individual NRSA that is awarded to an individual to specifically fund their training (the F31 and F32). In general, the term NRSA is used to refer to the individual grants, with the institutional grants referred to as T32s or simply training grants.

NSF – National Science Foundation (https://www.nsf.gov/). The National Science Foundation is an agency of the United States federal government that funds scientific research. This research spans fields of scientific inquiry, for example, physics and math, biological sciences, and engineering. This agency funds about 25% of all research at United States universities.

PI – Principle Investigator. This is the lead researcher and holder of a research grant whom assumes responsibility for the conduct of the research. This term is also used more broadly to refer to the head of a laboratory, even one without such grant funding.

Postdoc – Postdoctoral positions are those that people take after completing their PhD, but before beginning a more permanent position. These jobs typically focus almost exclusively on research, as the salary is paid through grants or other institutional research funds. However, some individuals do some teaching in these positions too.

R01 – This is the most common category of grant from the National Institutes of Health that funds laboratories through support of specific research projects described in the grant proposals. These grants last up to 5 years, and can have large budgets, although most are for up to $250,000 USD per year of direct research funding.

R1 – Research 1. This means the universities and colleges in the United States that grant PhDs and have the highest levels of research activity. The term comes from work by the Carnegie Foundation in 1970 to classify and quantitatively rank universities in the United States (http://carnegieclassifications.iu.edu/classification_descriptions/basic.php).

RA – Research Assistant. These are laboratory jobs. Paid RA positions are those in which people work in exchange for money. These positions are typically filled by people after they have graduated from college but before attending graduate school. These are sometimes known as post-baccalaureate jobs, to distinguish them from unpaid RA positions in which undergraduates work in laboratories in exchange for credits toward graduation, or sometimes in exchange for the research experience itself. The latter type of RA position is similar to an intern position at a company.

SLAC – Small Liberal Arts College. This describes a group of colleges which focus on undergraduate education, with the goal of giving their students a broad education to develop general thinking skills.

Startup – These are funds that a university grants to new tenure-track faculty members to start up their research. The idea is that this money gets your lab set up and gets your research far enough along that funding from grants can take over the support of your research program.

TA – Teaching Assistant. This is a position that assist professors with teaching a class. Typically, these are graduate students who receive their stipend in exchange for this service in the classroom. These positions typically involve grading, holding office hours, running discussion sections or review sessions, and sometimes lecturing. Some universities have undergraduate TA positions in which college students perform these tasks after having taken the class and performed particularly well.

Tenure – This means to give someone a permanent job. Being granted tenure means greater job security and typically a raise in pay. At universities in the United States it also typically comes with additional responsibility in that you are expected to help run the department and university by holding service positions (director of graduate studies, chair, university-wide committee member, et cetera).

Tenure track – These are faculty jobs in which the faculty member is evaluated for tenure 4–10 years after beginning hired as an assistant professor. This term distinguishes these positions from adjunct or lecturer positions that employ PhDs to teach on short-term contracts (usually 1–3 year contracts) at universities in the United States.

Acknowledgments

We are grateful to the reviewers of this book who provided thoughtful feedback on a portion of the chapters, including Alita J. Cousins, PhD, Eastern Connecticut State University; Brian L. Keeley, PhD, Pitzer University; Evangelia G. Chrysikou, PhD, Drexel University; and Chip Folk, PhD, Villanova University. We especially thank Molly Erickson, PhD, who reviewed the entire book and answered numerous follow-up questions, with humor, eloquence, and insight.

We thank the folks at Cognella for their support in our creation of this project. In particular, nobody returns emails as quickly and knows their projects as intimately as our fantastic Project Editor, Michelle Piehl, and Acquisitions Editor, Jen Codner. Emely Villavicencio, Senior Graphic Designer, makes the most gorgeous book covers and patiently entertains indecisive authors.

We are thankful for conversations with Missy Beers, PhD, about the importance of teaching in higher education and the impressive program she runs in the psychology department at The Ohio State University to train graduate students how to teach. We thank the undergraduate members of the Maxcey Lab for thoughtful comments on earlier drafts of chapters in this book and graduate students and postdocs of the Woodman Lab for their comments.

We are thankful to Hunter, Sam, Paige, and Henry who serve as constant reminders that we are working with other people's children, encouraging us to mentor our students and advisees as thoughtfully as we hope someone will mentor our children someday.

The Conversation Continues on Brain Bios Podcast

brainbiospodcast.com

As we finished this book we realized we were not done with the topic. Please join us on iTunes where we continue the conversation on the Brain Bios Podcast.

Find us on iTunes

Index

A

accredited internship, 8
adaptive controls, 204
administrator issues, 172–173
American Psychological Association (APA) accredited internship, 8
Animal Care and Use Committee, 176
APA dissertation funding, 78
application fees, 25
apprenticeship, 20
authorship, 15, 58, 156–158

B

best-case scenario, 15–16
birthday gifts, 25
Boroditsky, Lera, 138–147
   dealing with two-body problem, 139–141
   grant application process, 146–147
   job talk tips, 143–144
   mentoring statement, 141
   negotiation of salary, 146–147
   proactive framing, 145–146
   teaching experience, 142–143
Brockmole, James, 29
Buffalo, Elizabeth
   awards and recognition, 91
   on postdocs, 93–94
   postdoctoral experience at NIH, 91–92

C

campus interview, 109–111
campus interviews, 121–122
candy-bowl professors, 150
career
   development, 71–74
   in academia, 2
chalk talk, 134
Chronicle of Higher Education, 105
clinical PhD program, 8
collaborations, 186–188
Computer Programming for Psychology Majors, 227
conflict, student-advisor, 76–77
cover letter, 104, 115–116
18 credit-hour semester, 9, 30
Culham, Jody, 70
curriculum vitae (CV), 14, 68

D

data collection and analysis, 61–63, 68
deferral, 135
D'Esposito, Mark, 210–219
developmental science, 176
diversity
   statement, 107–108
   steps to increase, 186–188, 198–199
doctor of psychology (PsyD) programs, 7
   advantages of, 8–9
   cost for, 9–10
   internship match rates, 8
   potential strengths of, 9
double majoring, 11–12

E

elite private school, 50
email address, 71
European Union (EU) grant, 87
external letter writers, 176
extracurricular activities, 12–13

F

faculty jobs, 2
faculty mentors, 14, 59, 87
   contacting potential, 19–22
first pre-tenure review, 192
F.O.C.U.S. program (Families of Children Under Supervision), 18
formative graduate school experience, 75
friends in graduate school, 73
full-time paid research assistants, 230–231

G

gift ideas, 25
Gmail, 71
Golomb, Julie, 52–56
good life checklist, 256
grade point average (GPA) scores, 6, 9, 26
   getting a 4.0, 28
   importance of, 9–11
   undergraduate course selection and, 29–31
grading scale, 58
graduate mentors, 33
graduate programs
   application fees, 25
   at Ohio State, 101
   benefit of preselecting of schools, 25
   chances for getting into, 26
   common mistakes to avoid, 34–35
   criteria to admittance, 20
   deciding on, 53–54
   default attitude of students, 75
   distinguishing between undergrad and, 77
   goal of, 57
   offer process, 47–48
   selecting, 34
   talking to current graduate students, 34
graduate record examination (GRE) scores, 6, 9, 20, 26, 31
   importance of, 9–11
   preparing for, 17
graduate recruiting, 26
graduate school, choosing
   choosing mentor, 48
   declining offer, 51
   evaluating feedback of mentors, 50–51
   location, 49–50
   name-brand schools, 50
   stipend amount, 49–50
graduate school mentors, 9
grants, 109
   application, 28
   for postdocs, 86–87
   for undergraduate-serving institutions, 208–209
   funding, 40–41
   getting, 86–90
   mentor's, 87
   NSF fellowship, 87
   proposal, 177–178, 202–203
   review committee, 78
   reviewers of, 200–202
grant writing process, 78–79, 90, 204, 210
   alternate outcomes section, 211
   establishing as expert, 216
   for undergraduate-serving institutions, 208–209

pilot data, 211–212
program officials, role of, 209–210
proposing hypotheses, 203–204
revising approach, 219
specific aims page, 212, 216–217
using figures, 204, 211–212
voice of narrative, 214–215
group therapy, 18

H

Hecht, Lauren, 115–124
higher education, 1
Hilverman, Caitlin (Hilliard), 253–257
historically black colleges and universities (HBCU), 107–108
holidays, 25
human cognitive neuroscience, 22
  lab, 86

I

imposter syndrome, 60
independent research scientist, 58
independent scientist, becoming, 38
Institutional Review Board, 176
institution-specific research plans, 103–104
internship match, 8
internships, 8, 18
  accredited, 8
  experiences, 18
interviewing for graduate school
  choice of graduate schools, 39–40
  dress code for, 43
  faculty perspective, 53–54
  preparing for meetings, 42–43
  schedules, 44–45
  timing and travel, 41–42
intimate relationships, 73

J

Janterm, 10
Jiang, Yuhong, 74–79
job advertisements, 127
job talk, 130–131
juvenile delinquents, program for, 18

K

K99/R00 grant, 89
Kaas, Jon, 177

L

lab memory, 225–226
leadership position, 13
letters of recommendation, 23–24, 32, 98
  common errors in, 24
liberal arts colleges, 97–98, 109, 112
life after graduate school, 79
Logan, Gordon, 52

M

master's in social work (MSW), 7
master's programs, 5, 101
  characterization of, 6
  duration, 6
  fees, 6
  in psychology program, 7
  research interests, 37
  terminal master's degrees, 6, 70
  vs PhD program, 5–7
Maymester, 10
McNamara, Timothy, 190–198
meetings, preparing for, 42–43
mentors, 9, 83, 102
  advantages of junior, 54
  collaboration with, 186–188

evaluating feedback of, 50–51
graduate, 34
in senior colleagues, 194–195
junior vs senior, 54–55
personal reputation, 41
postdoctoral, 88
publication rates, 40
publication record, 40
questions to ask potential, 55
ranking of, 40–41
risks of working with, 55
training scheme, 46
mentorship
advising trainees, 182–184
apprenticeship model of mentoring, 46
mentoring styles, 45–46, 182
model, 37
receiving, 21–22
Midstates Consortium for Math and Science, 121
misconduct, allegations of, 74
misrepresentation of qualifications, 46–47

N

name-brand schools, 50
National Institutes of Health (NIH), 89
 Beth's postdoctoral experience at, 91–92
 grants, 89, 199–200, 202, 206, 210, 219
National Research Council, 26
National Research Service Award (NRSA), 87
National Science Foundation (NSF) program, 17
 fellowship grant application, 28
 funding percentages, 199
 grants, 219
Natural Science and Engineering Research Council (NSERC), 87

negotiation, 112–113, 124
 for R1 faculty jobs, 146–148
 salary, 136, 246–249
neuroscience programs
 admission decisions for, 23–24
 course requirements, 22–23
 differences between graduate programs in psychology and, 22–23
nonacademic jobs, 2
non-tenure-track jobs, 113–114, 149

O

offer process, 47–48
Ohio State University (OSU), 17
on-campus interview, 109
online teaching, 101

P

packet (package), 175
paid research assistantships, 26–27
pedagogical journals, 105
pedagogy workshops, 106
persistence, 75
personal statements, 18–19, 106
PhD programs, 5, 100, 108
 clinical, 8
 internships, 8
 research programs, 17
 stipend for, 6
 vs master's programs, 5–7
postdoctoral mentor, 87–88
 conversation with, 88–89
postdoctoral researcher (postdoc), 57
 advantage of, 95
 average salary, 81, 83
 benefits in lab, 94–95
 competitiveness of, 81–82
 contract duration, 81

definition, 81
expectations for independence, 95
faculty jobs, 81–82
finding a, 83–84
funding, 81
getting grants, 86–87, 89–90
learning methods, 86
mentoring styles, 87–88
opportunities for, 94, 96
post job ads for, 83
publishing findings, 90
scientific writing, 85–86
writing skill, 85–86
pre-tenure review, 192–193
Principle Investigator (PI), 200
programming skills, 15, 27, 43, 45, 48, 59, 61–62, 68, 76, 157, 183–184, 224, 226, 233
program official, 209–210
promotion of women and other minorities, 197–198
proposal, grant, 177–178, 202–203
Psi Chi Journal of Psychological Research, 15
psychology program vs neuroscience program, 22–23
publication record of mentor's lab, 40

Q

quiting graduate school, 69–71

R

recommendation letters, 23–24, 32, 98
relationships, developing, 73
Research Experience for Undergraduates (REU), 17
research experiences, 13
  at mentor's lab, 37
  best-case scenario, 15–16
  funding, 16–17
  paid summer research programs, 17
  publishing, 15–16
research-focused conversations, 73
researching the faculty, 39
research-intensive school, 112
research practicum, 15
research project
  data collection and analysis, 61–63, 68
  developing own, 60
  maintaining multiple projects, 63–65
research statement, 129
research statements, 108–109
resubmission of grant proposal, 206–207
R factor rankings, 26
R1 faculty jobs
  acceptance of, 137–138
  interview for, 131–134
  job advertisements, 127
  job talk, 130–131
  negotiating offer, 134–137
  parts of application, 128–130
  reasons, 125–126
  teaching expectations, 193–194
R1 faculty jobs
  chalk talk, 134
R01 grant, 89
rigor, 171–172
R1 universities, 125–126

S

schedules, 44–45
  data collection and analysis, 68
schizophrenia, 204
  adaptive control in, 204
  cognitive deficits in, 204
  EEG oscillations, 204
  psychotic symptoms of, 204

science career, 71
scientifically-oriented clinical programs, 19
scientific writing, 68, 77–78, 84–85, 180
  teaching, 180–182
scientist-mentor, 37
search committee, 110–111
second pre-tenure review, 192–193
sexual harassment, allegations of, 74
Sharfman, Glenn, 100, 149, 151, 160–173
Simons collaboration, 94
Skype conversations, 42
small liberal arts college (SLAC), 97, 222
  running a lab at, 226–227
startup costs, 103
startup funds, 103
statements
  diversity, 107–108
  mission, 109
  personal, 18–19, 32, 106
  teaching, 104–107, 122, 129–130
STEM (science, technology, engineering, and math) teaching discussions, 106
student-advisor conflict, 76–77
study section members, 200
succeeding at graduate school, 57
  common stumbling blocks, 75–76
  day one activities, 59
  developing own experimental design, 60–61
  grading scale, 58
  importance of persistence, 75
  measuring, 58–59
  rules of thumb on giving talks, 66–67
  time management, 67–69, 77–78
  volunteering, 65–66
succeeding at teaching institution
  authorship issues, 156–158
  boundary issues, 158
  of candy-bowl professors, 150–151
  research program, 155–158
  teaching evaluations, 151–154
  tenure expectations, 159–160
  time management, 154–155
successful lab at R1 institution
  collaboration with SLAC, 222
  during summer months or winter break, 224
  full-time paid research assistants, 230–231
  networking at conferences, 223
  using summer research grants, 223
  with collaborators, 222, 224
  with postdocs, 231
  with students with master's degree, 222–223
  with undergraduate RAs, 225, 227–228, 232
  with visiting guests, 231–232
summer research programs, 17
summer session, 9–10
Summer Undergraduate Psychology Research Experience Grants, 17
systems neuroscience, 22

T

talks, rules of thumb on giving, 66–67
teaching assistant (TA), 47
teaching evaluations, 151–152
teaching expectations at R1 institutions, 193–194
teaching-focused careers, 1
teaching job, 97
  as associate professor, 113
  at teaching-heavy institution, 98
  cautions while applying, 122–123
  classroom range, 111–112
  definition, 97

evidence of teaching effectiveness, 101
getting teaching experience, 100–102
interviewing, 109–111
mentoring of students, 102
negotiation, 112–113
of postdocs, 102–103
preparing application for, 103–113
suggestions on getting teaching experience, 115–124
teaching experience and, 100–101
Teaching of Psychology, 105
teaching statement, 104–107, 122, 129–130
teaching students, 190–191
   lecture slides, 189
   undergraduate classes, 189
   voting in tenure situations, 196
tenure-track faculty position, 98, 102, 113–114, 120, 135, 149
tenure-track job
   administrative perspective, 191
   advising trainees, 182–184
   collaborative project, 187–188
   grant proposal, 177–178
   prepare for lecture, 189–190
   research, importance of, 176–177
   tenure decisions, 175, 194
   time management, 178–180
   training students, 184–186
terminal master's degrees, 6, 70
time management, 67–69, 77–78, 154–155, 178–180

U

undergraduate career, 5
undergraduate course selection, 29–31
undergraduate research experiences, 13–15, 31–33
   funding, 33

undergraduate school, 77
University Office of Affirmative Action, 74

V

value added, 171–172
Vanderbilt Center for Teaching, 191–194
video conferencing, 42
volunteering, 56, 65–66

W

webpage, creating, 72
well-funded lab, 41
Wolfe, Jeremy, 229–236
working from home, 250–251
work-life balance, 237–238
   case of single parents, 245–246
   contribution to department's, 251–253
   dealing with infrequent feedback, 243
   family time, 243–245
   good life checklist, 256
   identifying best thinking time, 240
   managing email, 242
   maximising daily rhythm, 238–239
   organizing to-do lists, 239, 256
   rejecting requests, 250–251
   scheduling meetings, 243
   self-care, 256
   setting boundaries at home, 253–255
   time management, 240–241
writing
   grant, 78–79
   scientific, 68, 77–78, 180
   skill of postdoc researchers, 85

Printed by Libri Plureos GmbH in Hamburg, Germany